Amino Acid and Peptide Synthesis

John Jones

The Dyson Perrins Laboratory and Balliol College, University of Oxford

Second edition

OXFORD

UNIVERSITY PRESS

OXFORD
University Press
Great Clarendon Street, Oxford OX2 6DP

Oxford University Press is a department of the University of Oxford.
It furthers the University's objective of excellence in research, scholarship,
and education by publishing worldwide in

Oxford New York

Auckland Bangkok Buenos Aires Cape Town Chennai
Dar es Salaam Delhi Hong Kong Istanbul Karachi Kolkata
Kuala Lumpur Madrid Melbourne Mexico City Mumbai Nairobi
São Paulo Shanghai Singapore Taipei Tokyo Toronto

and an associated company in Berlin

Oxford is a registered trade mark of Oxford University Press
in the UK and in certain other countries

Published in the United States
by Oxford University Press Inc., New York

First Edition 1992
Second Edition 2002

A catalogue record for this book is available from the British Library
Library of Congress Cataloging in Publication Data

Data applied for

ISBN 0 19 925738 8

Typeset by EXPO Holdings,
Malaysia

Printed in Great Britain by
The Bath Press, Bath

Acknowledgement

Although the coverage of amino acid synthesis was new, the greater part of the 1992 printing of this primer was a distillate of the chapters on fundamentals which appeared in my book *The chemical synthesis of peptides* (OUP, 1991, reprinted in paperback 1994). I remain grateful for permission to use this material, but ten years have passed, and there have been three reprints in the meantime, each with additions and corrections. There are further extensive additions and adjustments in this second edition of the primer, which now supersedes my 1991 book altogether in giving, albeit more briefly, my view of the chemical essentials of mainstream peptide synthesis today.

J. H. J.
Oxford
February 2002

Contents

1   Introduction

Proteins are natural polymers which are assembled under nucleic acid control from a menu comprising L-α-amino acids of general structure **1** and the 'imino' acid L-proline (**2**): see Table 1.1. Amide or 'peptide'

(**1**)   (**2**)

Table 1.1   The common proteinogenic amino acids (1)

Amino acid*	$-$R
Alanine	$-CH_3$
Arginine	$-(CH_2)_3NHC(=NH)NH_2$
Asparagine	$-CH_2CONH_2$
Aspartic acid	$-CH_2CO_2H$
Cysteine	$-CH_2SH$
Glutamine	$-(CH_2)_2CONH_2$
Glutamic acid	$-(CH_2)_2CO_2H$
Glycine	$-H$
Histidine	$-CH_2(4\text{-imidazolyl})$
Isoleucine**	$-CH(CH_3)CH_2CH_3$
Leucine	$-CH_2CH(CH_3)_2$
Lysine	$-(CH_2)_4NH_2$
Methionine	$-(CH_2)_2SCH_3$
Phenylalanine	$-CH_2Ph$
Serine	$-CH_2OH$
Threonine***	$-CH(CH_3)OH$
Tryptophan	$-CH_2(3\text{-indolyl})$
Tyrosine	$-CH_2(4\text{-hydroxyphenyl})$ ˙
Valine	$-CH(CH_3)_2$

*All the proteinogenic amino acids belong to the same, L-, stereochemical series, and all save one are designated (2S); the way the Cahn–Ingold–Prelog sequence rules work makes L-[but (2R)-]-cysteine an apparent anomaly. This minor confusion has reinforced innate conservatism, and so far ensured the survival of the L-D terminology in the field.
**(2S, 3S)
***(2S, 3R)

Some specialized proteins contain residues which are not in the Table. Several involved in blood clotting and calcium metabolism, for example, contain γ-carboxyglutamic acid (Gla) residues, with R $= -CH_2CH(CO_2H)_2$, which bind Ca^{2+} tightly. As with most 'unusual' amino acids found in proteins, the unusual feature is introduced after backbone assembly from 'normal' proteinogenic amino acids. But at least one rare protein amino acid, selenocysteine, is incorporated as such under nucleic acid control.

L-[2S,3S]-isoleucine   L-[2S,3R]-threonine

The non-proteinogenic diastereoisomers with the opposite configurations at C-3 are called L-allo-threonine and L-allo-isoleucine respectively.

bonds link the building blocks, giving macromolecules (**3**) with

(**3**)

'polypeptide' backbones and side-chains containing a variety of simple functionalities according to the amino acids selected. These structures may be further elaborated by inter- or intra-chain covalent connection (most commonly by the formation of disulphide bridges between thiol side-chains, as in **4**); or by non-covalent association; or by the

CH₂SH   HSCH₂

oxidation

Disulphide bridge formation

Vasopressin   (**4**)

coordination of metal ions; or by the attachment of auxiliary components such as haem, carbohydrates, phosphate groups, or lipids; or by chemical modification in other ways, which include acetylation, hydroxylation, methylation, and carboxylation. Proteins may be acidic, basic, or neutral. They may be globular and soluble, or fibrous and insoluble. Their molecular weights span a range which extends from a few thousand to several million. They are present in abundance and diversity in every part of every living on Earth. Their indispensability was recognized long before anything about them could be understood in present-day terms, and is the origin of the name 'protein', which is derived from the Greek προτειοσ, ranked first.

Peptides are molecules constructed in essentially the same way as proteins, but on a smaller scale. The terms 'peptide' and 'protein' are used very inconsistently in the literature of the subject. 'Protein' is appropriate in current usage for any macromolecule incorporating more than about fifty of the usual α-amino residues assembled under nucleic-acid direction, or for a close synthetic analogue. The classification 'peptide' is suitable for molecules with fewer amino acid residues than this, whatever the amino acids and however produced, even if there is extensive modification after assembly of the residues by peptide bond formation. A prefix is often applied to indicate the number of residues— e.g. tripeptide, three residues.

The full structural formula of vasopressin (**4**), a pituitary nonapeptide which was one of the first of the peptide hormones to be synthesized (1954), and which is, from today's vantage point, a very simple example, is too complex to be taken in at a glance. A system of highly abbreviated formulae has been developed to present such structures. In this system, each amino acid has a three-letter code, derived from its trivial name. A one-letter code has also been agreed for the proteinogenic amino acids (i.e. those found in proteins), but it is more useful in connection with amino acid sequence data than for peptide syntheses. The simplification which can be achieved with the three-letter code is well illustrated by comparison of the full (**4**) and abbreviated (**5**) formulae for vasopressin. Together with cryptic symbols for substituents, the three-letter codes can be used to formulate any amino acid or peptide derivative, including synthetic intermediates. An International Commission has recommended detailed conventions which are, with minor aberrations, universally observed. Their system, which will be used throughout this book, is outlined in Appendix A, along with lists of amino acid codes and substituent symbols. Fluency with the system will be assumed, and readers are advised to study Appendix A before going further. The abbreviations used for common reagents and solvents are also defined there.

Catalytic proteins—enzymes—mediate practically all the operations in the molecular business of biology. Biochemical balance is also regulated by hormones, the majority of which are proteins or peptides. Membranes contain proteins which control permeability, and help to pump solutes through against thermodynamic gradients. Many neuroactive peptides and proteins with complex interrelationships have been identified in the brain. Selectively toxic peptides and proteins are deployed by many species, either for defence against predators, or in aggressive competition for limited nutrient resources. The venoms of snakes, bees, and wasps, for example, are evil cocktails of this sort; the lethal bacterial toxin responsible for botulism is a protein of high molecular weight; the principal poisonous constituents of the notorious toadstool *Amanita phalloides* are complex cyclic peptides; and antibiotics such as penicillin are, though modified almost beyond recognition as such, nevertheless peptides of a kind. The toxic peptides of bacteria and fungi are of special chemical interest because they are structurally very varied, incorporating D- as well as L-configurations, and amino acids which are often quite different from those of proteins. Carrier proteins transport smaller molecules from one location to another. Haemoglobin, loading and unloading oxygen as required, is a familiar example. Skin, bone, hair, horn, tendon, muscle, feather, tooth, and nail are all largely protein-aceous, as are spider's webs, larval cocoons, and antarctic fish antifreeze agents. The immune system, with which animals defend themselves against invasive parasites, employs proteins to recognize and reject foreign molecules, exercising exquisitely precise discrimination between self and not-self. Even genetics requires proteins, because although the genetic information is encoded in nucleic acid structures, the DNA is highly organized in compact association with basic proteins called

$\overline{\text{H·Cys·Tyr·Phe·Gln·Asn·Cys}}$·Pro·Arg·Gly·NH$_2$

Vasopressin (**5**)

Actually there are many penicillins, of the general structure shown below. The bicyclic system is biosynthetically derived from the dipeptide unit whose backbone is emboldened.

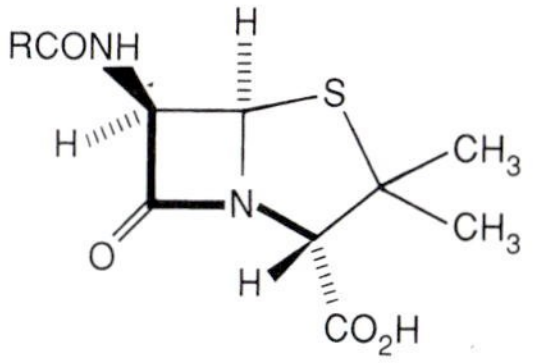

That the 3D (tertiary) structure of a protein is determined by the amino acid sequence (primary structure) of the protein or a precursor is well proven (see Section 9.2.2.2), but it is not yet possible to predict tertiary structures from sequence data.

Examples of synthetic peptide drugs include Synacthen® (an adrenocorticotrophin fragment, for the therapy of rheumatic diseases, bronchial asthma, allergic disorders, skin conditions, etc); DDAVP® (a vasopressin analogue, for the treatment of *diabetes insipidus*); Sandimmun® (cyclosporin A, a cyclic peptide immunosuppressive drug used in organ transplantation); Calsynar® (salmon calcitonin, for the assessment and management of hypercalcaemia, Paget's disease, bone cancer, etc); and Zoladex® (an analogue of LH-RH, luteinising hormone releasing hormone, for the treatment of prostate cancer and endometriosis). Zoladex sales alone are variously estimated at £300 to £500 million a year. See the review by Loffet, details in Appendix B.

histones, and an array of other proteins is necessary for genetic expression and control. Furthermore, the genes dictate only protein amino acid sequences. These determine three-dimensional structure, from which all else flows. All the data necessary to specify that an organism will have the form and characteristics of an orchid, an oryx, or an orang-utan is written in the amino acid sequences of its proteins.

It follows that there are few aspects of biology which cannot be illuminated by experiments using peptides or proteins. Nor is the interest merely academic, for peptides and peptide-related structures can influence endocrine, neurological, immune, and enzymic processes with high specificity and prodigious potency. They therefore have varied applications or potential in medicine: in the regulation of fertility; the control of pain; the stimulation of growth; the therapy of cancer, cardiovascular problems, connective tissue diseases, digestive disorders, mental illness, and infections by pathogens. Several peptide drugs are in widespread clinical use, necessitating their manufacture by the kilo, and are big business. Such drugs all ultimately depend on the selective activation or inhibition of an existing part of the biochemical apparatus, by the specific recognition of, and interaction with, natural receptors or active sites—a universal theme of peptide and protein biochemistry. Then there is the prospect of synthetic peptide vaccines. If the immunologically significant parts of a protein can be mapped, then in principle they can be mimicked in a synthetic peptide vaccine, which can be used to immunize against the agent they belong to. This strategy, avoiding as it would the hazards and limitations of making and using vaccines in the traditional way, has already been shown to be feasible, and seems certain to become very important. And now it also begins to be possible to conceive a fundamentally new kind of approach to peptide and protein applications. Much remains to be discovered about the relationship between amino acid sequence, three-dimensional structure, and function, but an unrestrained imagination can look forward to the time when the chemist will be able to design and construct entirely unnatural peptides and proteins with predetermined novel properties.

The demand for peptides and proteins is thus enormous, and rising all the time. Natural sources can provide a great variety, sometimes freely and in quantity, but the range is limited to what happens to be there, and more often than not only miniscule amounts can be isolated. The manipulation of biological systems by genetic engineering is an increasingly sophisticated art, and will no doubt become the principal means of manufacture for many natural and unnatural amino acid sequences. For example, insulin, a mini-protein of two chains and about fifty amino acid residues, which is essential for the treatment of *diabetes mellitus*, was formerly obtained from beef or pig pancreatic tissue. Because insulins from these sources are not quite identical with human insulin, some patients become sensitized against what are, to their systems, foreign substances. The obvious solution is to administer the human hormone. This demand could never adequately be met from natural sources, but most of the human insulin needed is now being prepared commercially by recombinant DNA technology. Several other important peptides are being produced in the same way. However, each such instance requires developmental work, and the approach is at present restricted in the face of structural novelty. Chemical synthesis, on the other hand, is in principle applicable to any target. Since medicinal chemists commonly turn to unnatural analogues in search of metabolic stability, and the screening of large numbers of variants is generally necessary in the development of a new drug, chemical peptide synthesis is likely to remain an essential tool for the forseeable future. It can be estimated that over five thousand people are engaged full-time in such work worldwide at the present time.

The aim of this book is to outline the principal chemical methods which are available for the synthesis of α-amino acids and peptides. Ruthless selectivity will be necessary, and there will be no space at all for technical matters (a deficiency greatly regretted, for this is a field in which the advances of the last four decades have been dependent on technology as well as chemistry). Similarly, even a rudimentary coverage of principles will leave little room for real examples. The subject and the molecules it has tackled with success are just too big. Furthermore, we cannot in the space available stray far from the proteinogenic amino acids and 'ordinary' peptides, although the synthesis of peptides containing other amino acids and non-protein structural features is of very great importance.

The synthesis of 'peptidomimetic' drug candidates, which have overall functionality and shape related to peptide leads, but metabolically stable non-peptide backbones, is a diverse and important field, but is beyond our scope.

Nature synthesizes proteins by stepwise assembly from amino acid building blocks, and this is the way the chemist generally proceeds in the laboratory.

The formation of a dipeptide **6** having no side-chain functionality by the condensation of two amino acids with the elimination of water can be

$$H_2NCHR^1CO_2H + H_2NCHR^2CO_2H \longrightarrow H_2NCHR^1CONHCHR^2CO_2H$$

(6)

Scheme 1.1 Conditions: dehydration

As an exercise, devise a synthesis of **6** which does not use either amino acid, and consider its problems.

represented formally as in Scheme 1.1. It is clear that even if the chemistry of this conversion could be achieved, which in the naive manner shown it cannot, it would be ambiguous and uncontrolled. There being no means of distinguishing between the two amino groups or the two carboxy groups, the product **6** would be accompanied by the dipeptides **7–9**, together with the polycondensation products from all

$$H_2NCHR^1CONHCHR^1CO_2H \qquad H_2NCHR^2CONHCHR^1CO_2H$$

(7) (8)

$$H_2NCHR^2CONHCHR^2CO_2H$$

(9)

four possible dipeptides. The controlled synthesis of **6** would entail the blockade (or 'protection') of the amino group belonging to one component (the 'carboxy component') and the carboxy group of the other (the 'amino component'), followed by condensation (or 'coupling'), and finally removal of the blocking groups from the intermediate product **10** ('deprotection')—a minimum of four separate reactions (Scheme 1.2,

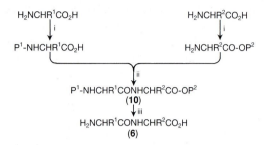

Scheme 1.2.   Conditions: i, protection; ii, coupling; iii double deprotection.

otherwise represented as in Scheme 1.3). For the conversion of **10** to a tripeptide **11**, there would be the additional requirement that P^1 should be removable without disturbing P^2, i.e. P^1 and P^2 should be 'orthogonal'. This would mean a minimum of eight separate reactions

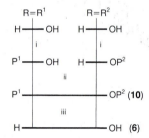

Scheme 1.3.   Conditions: i, protection; ii, coupling; iii, double deprotection.

for the synthesis of **11** via **10** (Scheme 1.4) from free amino acids. Then there is the complication that the functional side-chains of the

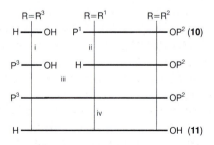

Scheme 1.4.   Conditions: i, protection; ii, selective deprotection; iii, coupling; iv, double deprotection.

proteinogenic amino acids all interfere with the reactions available for peptide bond formation to some degree, and usually require protection orthogonal to at least one of the α-function protection methods chosen, which limits choice and adds to the number of operations. An alternative approach is to separate the carboxy activation and coupling stages. In this case, the amino component may be attached by a 'salt coupling', without the need to protect its carboxy group, provided exchange of activation between the free and activated carboxy groups does not intervene (Scheme 1.5). This stratagem reduces the number of protecting

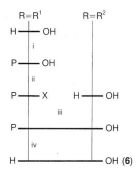

Scheme 1.5.   Conditions: i, protection; ii, activation; iii, coupling, with the amino component in the salt form; iv, deprotection.

group manipulations, but introduces an extra operation at the coupling stage, and may be disadvantageous for other reasons such as solubility.

Furthermore, all the transformations in a peptide synthesis must needs be performed without loss of stereochemical integrity at any of the chiral centres, if a product free of very similar, and possibly inseparable, diastereoisomeric contaminants is to be obtained. Lastly, the final removal of all the protecting groups must be carried out without destruction of the peptide backbone or other side-reactions. All of which makes the unambigious synthesis of large peptides sound unattainable. In practice things are not quite as bad as this. To start with, although unusual amino acids may have to be synthesized (Chapter 2) and appropriately derivatized (Chapters 3–5) before peptide synthesis can begin, the requisite derivatives of all the standard amino acids, protected and ready for incorporation, are commercially available. And the methodology of the subject has now been finely tuned, automated synthetic techniques have been developed, and modern purification technology can in fact cope with very complex mixtures.

It is convenient to discuss the chemistry and merits of the main kinds of protection and of the reactions available for peptide bond formation in turn (Chapters 3–6), but it must be understood that no rigid rules can be laid down prescribing the best weapons or plan of attack for any particular synthesis. The protecting groups and coupling procedures must be able to work together without conflict, and the overall strategy and orchestration (Chapters 7–9) require judicious planning just as much as the individual steps call for care in the selection of the optimal methods.

Furthermore, a synthetic route used for discovery and development might not be appropriate for large-scale production. See the review by Andersson *et al.*, details in Appendix B.

2   α-Amino acid synthesis

The proteinogenic α-amino acids are produced industrially by fermentation methods and by chemical synthesis on a vast scale, reckoned for some in hundreds of thousands of tons per annum. Their principal application is as food additives, but they are incidentally cheap starting materials for laboratory work. The synthesis of α-amino acids at the bench is nevertheless an active field because of the demand for specifically labelled, unnatural and unusual amino acids. The need is almost always for a homochiral product, so assembly of the target without regard to α-chirality must be followed by resolution; alternatively an asymmetric synthesis must be employed; or an enantiospecific conversion of a freely available homochiral compound to the required α-amino acid must be achieved. Examples of all three approaches will be given.

2.1   General synthetic methods

Many of the general methods for the synthesis of α-amino acids, including displacement reactions on α-halo acids (e.g. Scheme 2.1), the Strecker

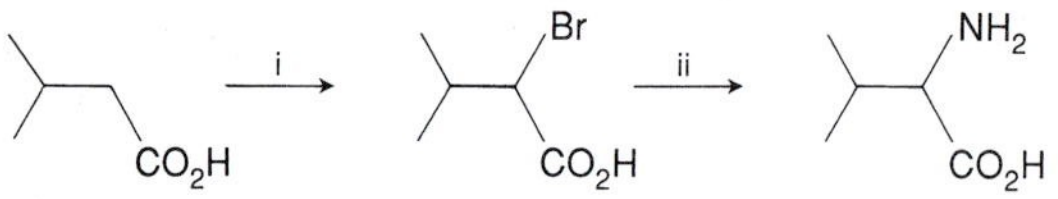

Stage i is the Hell–Volhard–Zelinsky reaction; it probably proceeds via the enol of the acyl halide, and thus only α-bromination takes place.

Scheme 2.1.   Conditions: i, Br_2/PCl_3; ii, NH_3.

synthesis (e.g. Scheme 2.2), approaches through hydantoins (e.g. Scheme 2.3), and via oxazolones (e.g. Scheme 2.4), were developed in the early days of amino acid chemistry, but retain their importance. Although ammonia works well enough in conversions of very simple α-halo acids to

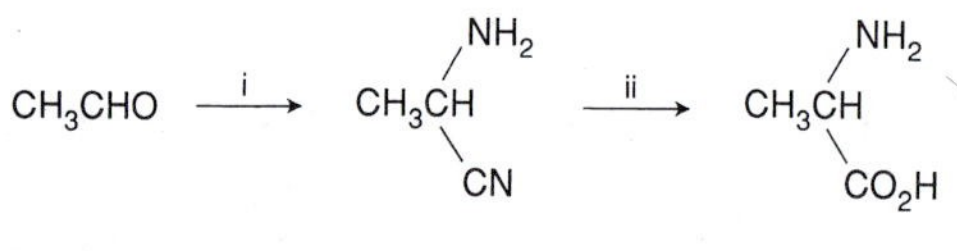

Scheme 2.2.   Conditions: i, $NaCN/NH_4Cl$; ii, H_3O^+.

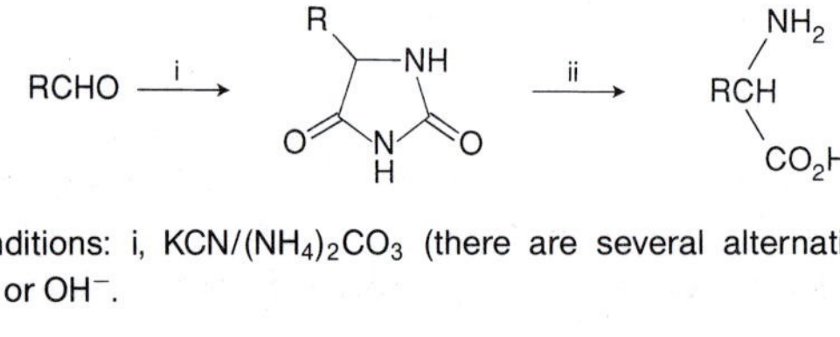

Problem: suggest a mechanism for stage i.

Scheme 2.3.   Conditions: i, $KCN/(NH_4)_2CO_3$ (there are several alternatives for this ring-formation); ii, H_3O^+ or OH^-.

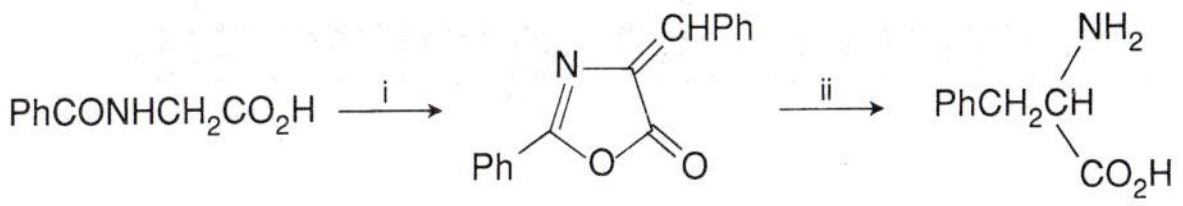

Scheme 2.4. Conditions: i, PhCHO/Ac$_2$O/NaOAc; ii, aq. HI/P/heat.

Stage i here is the Erlenmeyer azlactone synthesis (azlactone = oxazolone); it involves cyclodehydration of benzoylglycine to give an oxazolone, which has enhanced acidity at the CH$_2$ group and condenses easily with benzaldehyde.

α-amino acids, potassium phthalimide (followed by strong acid hydrolysis of the intermediate phthalimido derivative: the Gabriel synthesis) or azide ion (followed by reduction) are superior reagents.

However, none of the above procedures is as frequently employed as that involving acylaminomalonates (e.g. Scheme 2.5). Although amino-

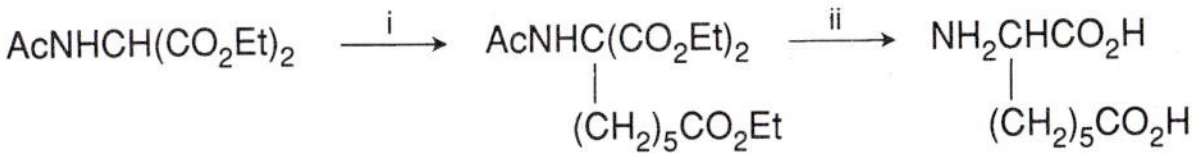

Scheme 2.5. Conditions: i, NaOEt, then Br(CH$_2$)$_5$CO$_2$Et; ii, H$_3$O$^+$.

malonic acid and its α-alkyl derivatives are isolable, they are unstable with respect to decarboxylation, so that α-amino acids are produced directly under the vigorous conditions of the last stage. Diethyl acetamidomalonate is easily obtained as in Scheme 2.6, and a range of other

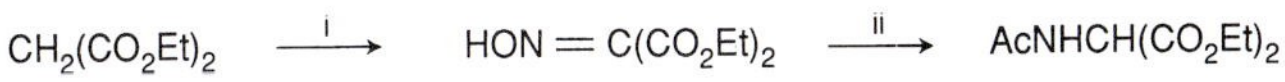

Scheme 2.6. Conditions: i, NaNO$_2$/AcOH; ii, H$_2$/Pd(C), then Ac$_2$O.

Problem: suggest a mechanism for stage i.

acylaminomalonates can be made and applied analogously (see Scheme 2.10).

Two further general strategies are illustrated in Schemes 2.7 and 2.8.

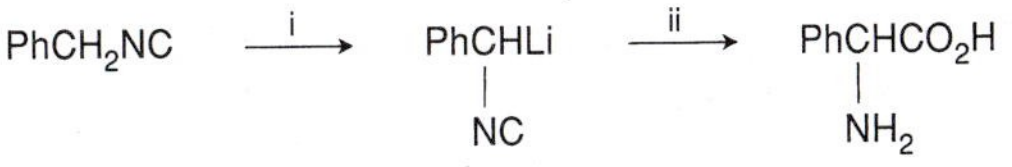

Scheme 2.7. Conditions: i, Bu^{n}Li; ii, CO$_2$ at −80 °C, then H$_3$O$^+$.

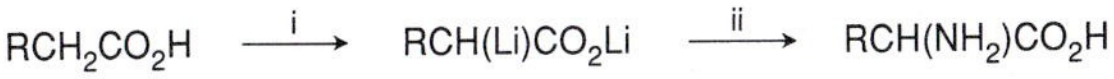

Scheme 2.8. Conditions: i, Pr$^i_2$NLi; ii, NH$_2$OMe.

This is an electrophilic amination, using nucleophilic C and electrophilic N for C–N formation. Most classical procedures for C–N formation involve nucleophilic N and electrophilic C.

2.2 Resolution

The reactions illustrated in Section 2.1 all inevitably lead to racemic products. The traditional general approach to the resolution of racemates of all kinds is to derivatize with an optically active reagent, separate the diastereoisomers, and then reverse the derivatization. Scheme 2.9 shows an example which works well with benzyloxycarbonylamino

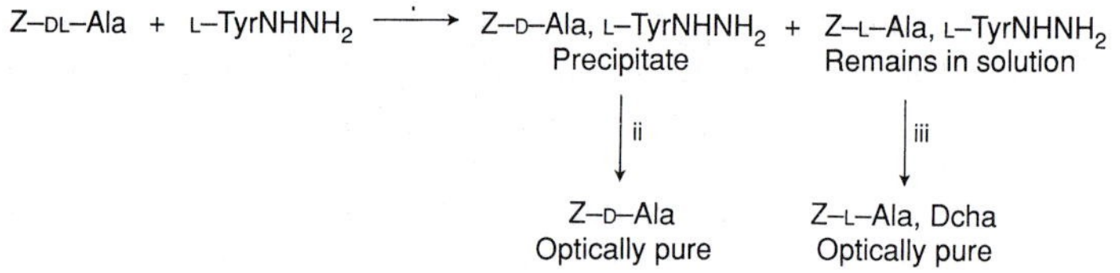

$$Z-DL-Ala \ + \ L-TyrNHNH_2 \ \longrightarrow \ \underset{\text{Precipitate}}{Z-D-Ala, \ L-TyrNHNH_2} \ + \ \underset{\text{Remains in solution}}{Z-L-Ala, \ L-TyrNHNH_2}$$

$$\underset{\text{Optically pure}}{Z-D-Ala} \qquad \underset{\text{Optically pure}}{Z-L-Ala, \ Dcha}$$

ii   iii

There is nothing especially subtle about the choice of dicyclohexylamine here; it just happens to form nice crystalline salts with a wide range of acids (cf. p. 21).

Scheme 2.9. Conditions: i, mix equimolar amounts in MeOH; ii, filter off, recrystallize once and neutralize (recovery of resolving agent); iii, neutralize liquors (recovery of resolving agent), add dicyclohexylamine (Dcha), and recrystallize once.

acids (see Section 3.1.1): the D–L and L–D diastereoisomeric salts differ so much in solubility and ease of crystallization that the former separates practically quantitatively. Enzymes provide an alternative approach which is especially appropriate for α-amino acids. If an enzyme catalyses a conversion at all, it is usually highly enantioselective, and a suitable enzyme and conversion can be found for most cases. Scheme 2.10 shows a convenient combination of the acylaminomalonate synthesis

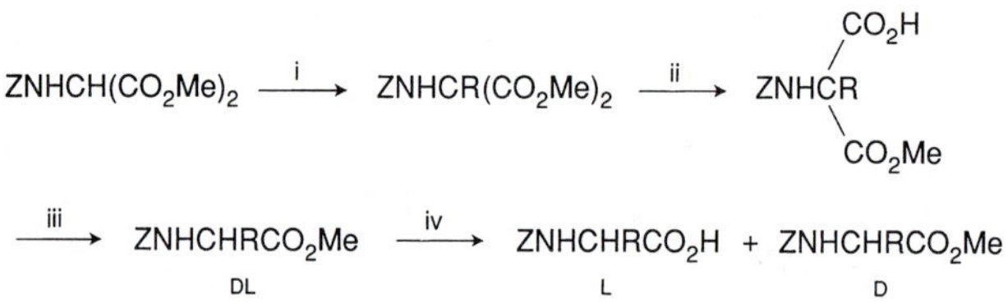

$$ZNHCH(CO_2Me)_2 \ \xrightarrow{i} \ ZNHCR(CO_2Me)_2 \ \xrightarrow{ii} \ ZNHCR\begin{smallmatrix}CO_2H\\ \\CO_2Me\end{smallmatrix}$$

$$\xrightarrow{iii} \ \underset{DL}{ZNHCHRCO_2Me} \ \xrightarrow{iv} \ \underset{L}{ZNHCHRCO_2H} \ + \ \underset{D}{ZNHCHRCO_2Me}$$

Scheme 2.10. Conditions: i, NaOEt, then RX; ii, KOH until one group saponified; iii, heat in dioxan; iv, protease (subtilisin *Carlsberg*).

with an enzymatic method of resolution; Scheme 2.11 outlines a different tactic with a different enzyme which is particularly valuable because

This enzyme is also called hog renal acylase.

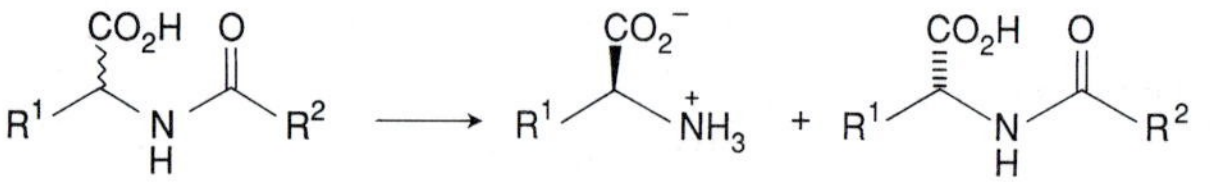

Scheme 2.11. Conditions: acylase I (aminoacylase) from porcine kidney.

although the enzyme used is enantioselective, it works with a wide range of side chains.

2.3 Asymmetric synthesis

For a reaction which creates a new chiral centre to lead to more of one form than the other, i.e. for asymmetric synthesis to take place, at least one of the reagents must be optically active, but very high enantiomeric excesses (ee) can sometimes be obtained. Many ingenious techniques have been devised for the efficient asymmetric synthesis of α-amino acids. Schemes 2.12–14 illustrate three of particular interest.

See Primer 63 for a general account of this topic.

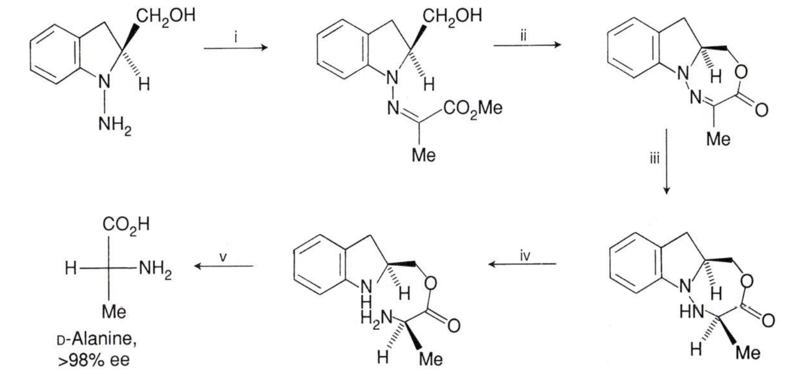

This is Corey's method. In stage iii, the new C–H bond is formed preferentially on the most accessible side.

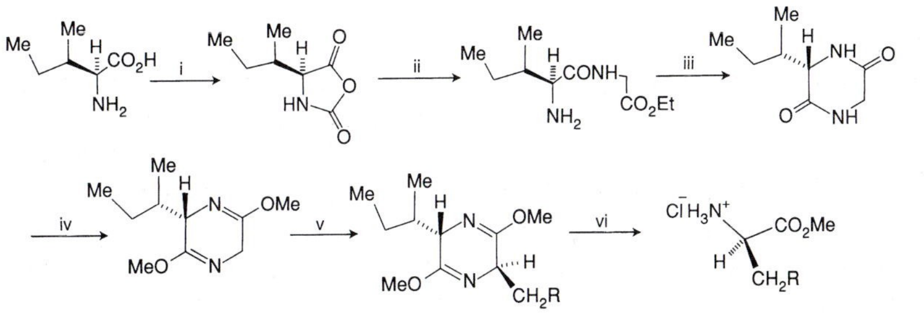

Scheme 2.12. Conditions: i, MeCOCO$_2$Me; ii, NaOMe/heat; iii, Al(Hg)/H$_2$O; iv, H$_2$/Pd(C), followed by recrystallization to remove traces of the minor diastereoisomer; v, H$_3$O$^+$.

Stage ii involves nucleophilic attack at the more electrophilic carbonyl of the N-carboxyanhydride, followed by spontaneous decarboxylation of the carbamic acid produced (cf. Scheme 5.27). Stage iii involves diketopiperazine formation (cf. Scheme 4.1). The product of stage iv is termed a bis-lactim ether. It forms an anion in stage v which is alkylated on its least hindered side. The whole approach is associated with Schöllkopf's name.

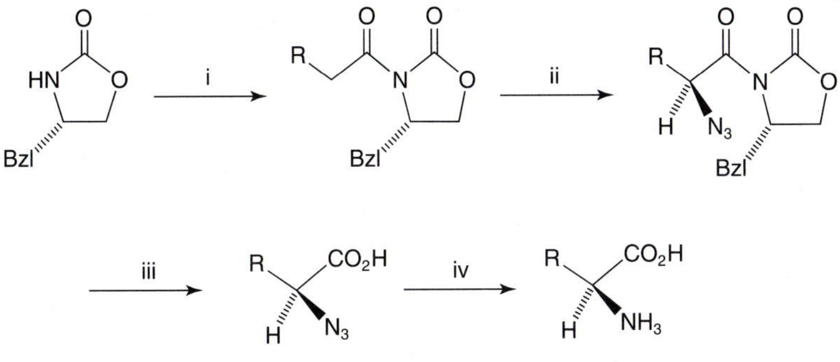

Scheme 2.13. Conditions: i, COCl$_2$; ii, NH$_2$CH$_2$CO$_2$Et; iii, heat; iv, Me$_3$O$^+$BF$_4^-$; v, Bu^{n}Li, then RCH$_2$X; vi, 0.25M HCl. Products with 70–95% ee were obtained, depending on R.

This chemistry was devised by Evans. How might the starting material be obtained from L-phenylalanine? The key step ii involves a chelated enolate reacting on its least hindered side with the electrophilic azide transfer reagent ArSO$_2$N$_3$.

Scheme 2.14. Conditions; i, BuLi, then RCH$_2$COX, low temp; ii, KN(SiMe$_3$)$_2$, then ArSO$_2$N$_3$, low temp; iii, LiOH ; iv, reduction. 99% ee.

2.4   Enantiospecific synthesis

In principle, the selective chemical modification of the side chains of cheap proteinogenic amino acids offers attractive access to rarer or unnatural L-α-amino acids, provided the original chirality can be preserved. L-Vinylglycine, for example, can be obtained from α-diprotected L-methionine or L-glutamic acid (Scheme 2.14).

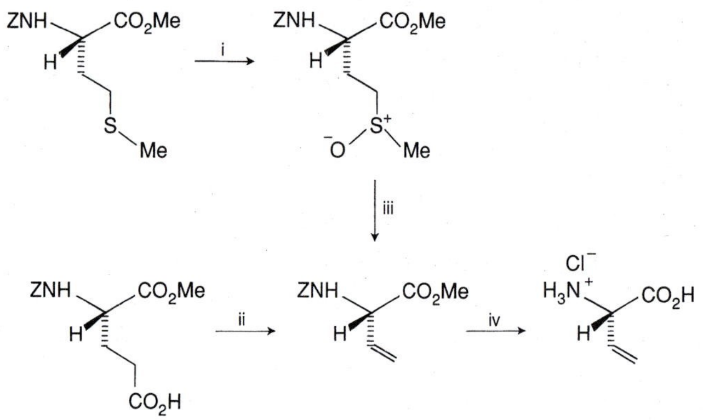

Scheme 2.15. Conditions: i, $NaIO_4/MeOH/H_2O/0\,^{\circ}C$; ii, $Pb(OAc)_4/Cu(OAc)_2/PhH/reflux$; iii, pyrolysis $148\,^{\circ}C/3$ mm; iv, 6N HCl/reflux.

Further reading

The synthesis of α-amino acids is a big field: see the appropriate books listed in Appendix B, and for an overview of classical procedures see the chapter on amino acids in *Comprehensive Organic Chemistry* (Barton, D.H.R. and Ollis, W.D., eds., 1979). The application of modern methods to the asymmetric synthesis of α-amino acids is of particular interest: for a useful summary see Dugas, H. (3rd edition, 1995) *Bio-organic Chemistry*, p. 51.

3 α-Amino protection

If an α-amino group is to be protected in the context of peptide synthesis, its nucleophilic reactivity must be suppressed, by draining its electron density away into an appropriate substituent, or by concealing it altogether behind a screen of gross steric hindrance. To be useful, protection according to one or both of these simple principles must be achieved easily at the outset; the protecting group should introduce no problems of its own while in position; it must stay firmly at its post as long as it is needed; and when its job is done, it should slip quietly away on command, under conditions which have no adverse effects on the rest of the structure being assembled—all without jeopardizing the chiral integrity of nearby chiral centres. Except in the special case where, at the end of a synthesis, several protected functionalities have to be exposed simultaneously to give the completed target molecule, there may be the additional requirement that the amino-protecting group should be orthogonal to the other protecting groups which are in play.

See Primer 95 for a broad-based account of protecting group chemistry.

These are demanding criteria, ruling out of serious consideration all simple alkanoyl groups, which in some respects might seem obvious candidates. Thus, nothing could be more straightforward than acetylation, which satisfies the fundamental requirement of suppressing nucleophilic reactivity very well, and is hydrolytically reversible. Unfortunately, the severe conditions required for hydrolytic deacetylation devastate most peptide chains too. This difficulty can be eased by building in the potential for mild intramolecular deprotection, as in Scheme 3.1, but the protecting group so derived is nevertheless still quite

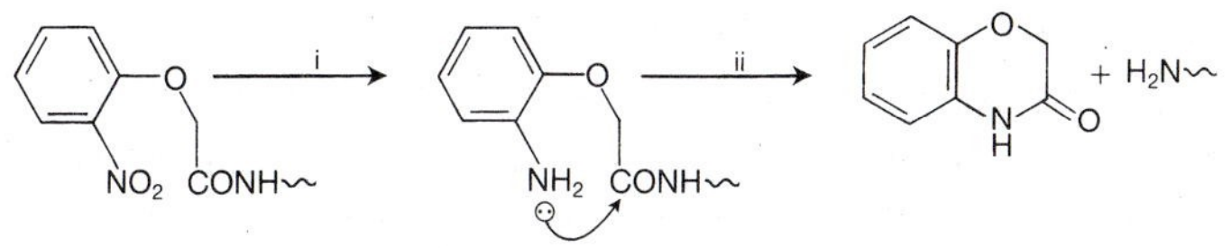

Scheme 3.1. Conditions: i, $H_2/PtO_2/aq. NaHCO_3$; ii, $H_2O/100\,°C/1\,h$.

In a related reaction, $ClCH_2CONH\sim$ releases $NH_2\sim$ on treatment with $(NH_2)_2C=S$: explain.

useless for α-amino protection in peptide synthesis on other counts, especially in relation to racemization. The benzoyl and 4-toluenesulphonyl ('tosyl') groups can be dismissed similarly—their introduction is trivial, and they are good electron-withdrawing groups, but they otherwise fail to fit the bill.

The development of α-amino-protecting groups which do meet the necessary specifications has entailed an extended investigation of a very large number of variations on a number of themes. Many subtly different groups have been proposed, but only those of real practical importance or promise will be reviewed.

3.1 Alkoxycarbonyl protection

Amino groups are easily converted to alkoxycarbonylamino derivatives **1** (otherwise 'urethanes' or 'carbamates'), which are both amides and esters. As amides, the derivatives **1** have low nucleophilic reactivity at

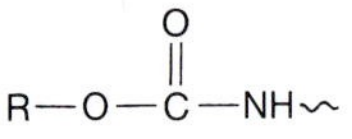

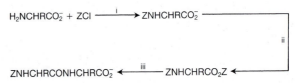

nitrogen. As esters, although they have relatively low reactivity with respect to acyl–oxygen fission, they can be degraded by alkyl–oxygen fission to the corresponding carbamic acids. Since carbamic acids decarboxylate spontaneously under practically all conditions, alkyl–oxygen fission regenerates the parent amine. Alkyl–oxygen fission can be achieved in a great variety of ways, depending on the alkyl group. Most of the amino-protecting groups of the first rank depend on these simple principles, and on the further vital fact that the alkoxycarbonyl derivatives of α-amino acids are very nearly immune to racemization (see Section 5.1.2).

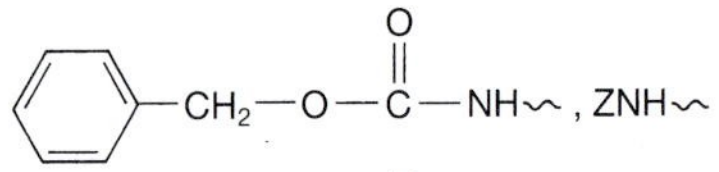

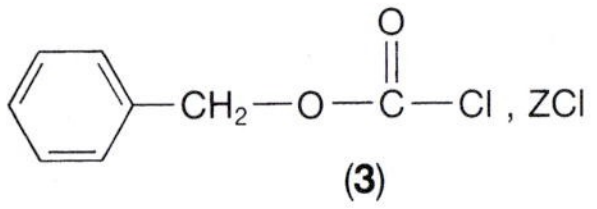

Carbamic acids decarboxylate easily.

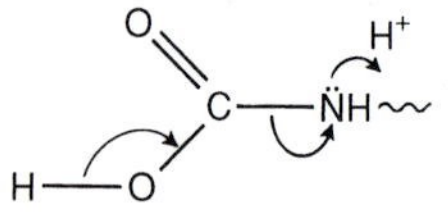

3.1.1 Benzyloxycarbonyl (Z) protection (2)

Bergmann and Zervas (from whose name the abbreviated designation Z derives) described this method of protecting amino groups in 1932. Their work is generally regarded as the turning point which marks the origin of modern peptide synthesis.

The Z group is still normally introduced with the original chloroformate reagent (**3**), but minor side reactions, such as dipeptide formation under Schotten–Baumann conditions with amino acids (Scheme 3.2), are

Chloroformates ROCOCl may be prepared by treating alcohols with a large excess of phosgene.

$$H_2NCHRCO_2^- + ZCl \xrightarrow{\quad i \quad} ZNHCHRCO_2^-$$

$$ZNHCHRCONHCHRCO_2^- \xleftarrow{\quad iii \quad} ZNHCHRCO_2Z \xleftarrow{\quad ii \quad}$$

Stage ii produces a mixed carbonic anhydride intermediate (cf. Scheme 5.12).

Scheme 3.2. Conditions: i, alkali (Schotten–Baumann); ii, reaction with unconsumed ZCl; iii, reaction with unconsumed $H_2NCHRCO_2H$.

occasionally a nuisance, and less reactive acylating agents, e.g. **4**, are sometimes to be preferred.

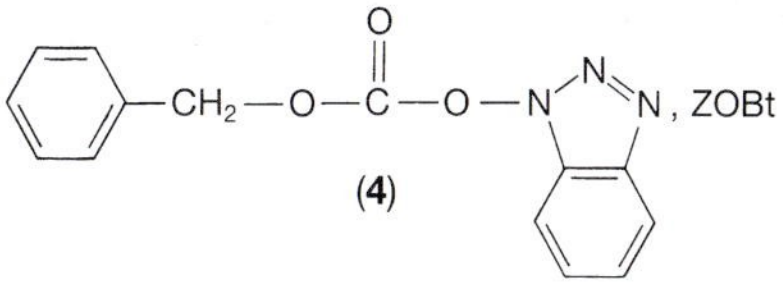

(4)

Mildly basic and or nucleophilic reagents—amines, hydrazine, dilute aqueous alkali—do not as a rule affect the Z group at ambient temperature, and the main exception (hydantoin formation, Scheme 3.3) is only

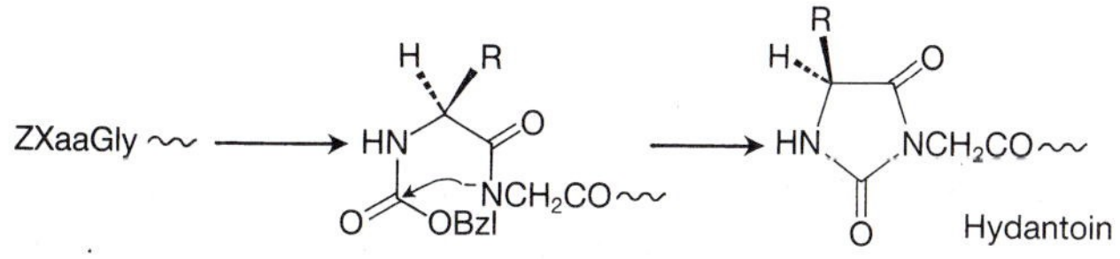

Scheme 3.3. Conditions: alkali.

Xaa is used here to designate any α-amino acid residue.

really significant with ZXaaGly-peptides. Mildly acidic conditions—of severity up to say half an hour's exposure to neat TFA in an ice-bath—are also without effect, and it is possible to carry a Z group unscathed through a wide range of functional group interconversion procedures, including many which are not met in routine peptide synthesis.

The classic cleavage conditions are HBr/AcOH or catalytic hydrogenolysis.

Cleavage by HBr/AcOH proceeds by an essentially S_N2 mechanism (Scheme 3.4) near the S_N1/S_N2 borderline. It is very convenient to

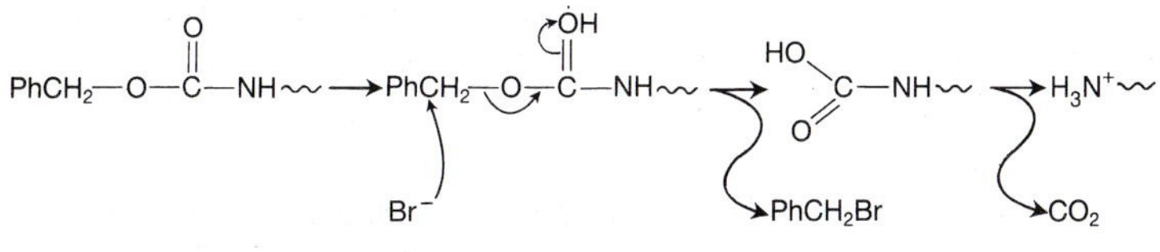

Scheme 3.4. Conditions: HBr/AcOH.

carry out, because most protected peptides are soluble with ease in acetic acid, especially in the presence of a high concentration of HBr, and the deprotected product may be precipitated as a hydrobromide salt on flooding with ether. HBr/AcOH is a highly corrosive system, and it is perhaps surprising that most simple peptides survive it, and other highly polar and acidic media, unharmed, but some do not. There are deleterious effects in more complicated cases, especially if certain susceptible side-chains are present (see Chapter 6). These problems are usually blamed on the generation of electrophilic species—benzyl cations or their equivalents—which can attack electron-rich side-chains (e.g. Trp, Tyr, Met). Additives like anisole and dimethyl sulphide diminish difficulties of this kind, by acting as 'scavengers' for such species. Exactly what 'scavenging' means in this context is more complicated than it sounds, because most of the popular scavengers are also soft nucleophiles. They can intervene not only simply by sequestering any electrophiles before havoc is caused, but also by themselves acting directly as nucleophiles in the cleavage stage itself, assisting it and smoothing its path by shifting its position in the mechanistic spectrum towards the S_N2 extreme. Increasing the acidity and or polarity accelerates the cleavage; liquid HF and HBr/TFA both remove Z groups rapidly. With these powerful reagents there is probably a more S_N1-like mechanism in operation, so problems from wandering electrophiles (which, furthermore, enjoy a longer lifetime because the solvent is less nucleophilic) are more serious, and scavenging additives assume a correspondingly more important role.

> Liquid HF, however, is an extremely hazardous and difficult-to-handle reagent.

A protic acid is not essential for cleavage. The necessary electronic demand can be provided by a Lewis acid too, as in deprotection with trimethylsilyl iodide (Scheme 3.5).

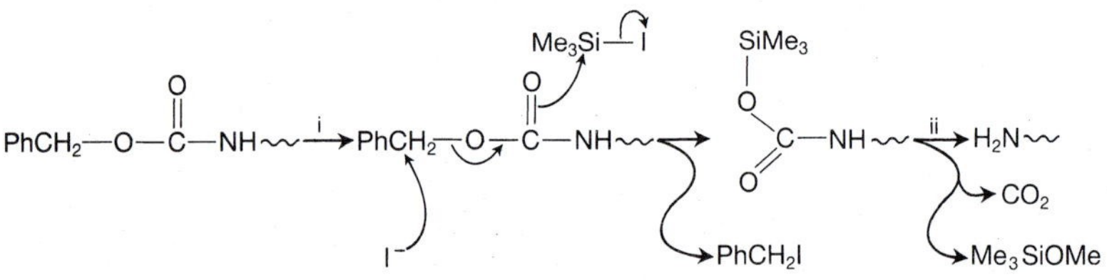

Scheme 3.5. Conditions: i, 1.2 equiv. Me$_3$SiI/MeCN or CHCl$_3$/25 °C/10 min.; ii, quench with MeOH.

Electron-releasing substituents in the aromatic ring affect the rate and mechanism of acidolysis, accelerating the cleavage and making it more S_N1-like; and electron-withdrawing groups do the opposite. The 4-methoxybenzyloxycarbonyl group (**5**), which is cleaved by TFA, is the best-known and most useful ring-substituted analogue.

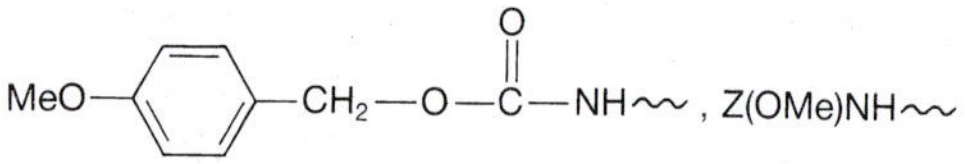

(**5**)

Catalytic hydrogenolysis (Scheme 3.6) of Z groups can be achieved with a variety of catalysts and conditions, of which 80% acetic acid/

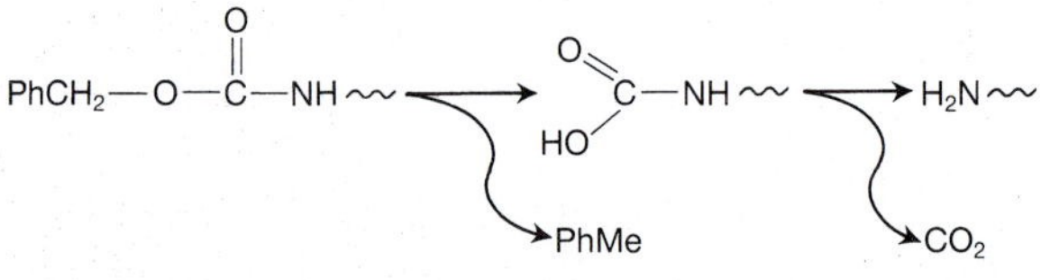

Scheme 3.6. Conditions: H$_2$/Pd(C)/aq. AcOH/1 atm./20 °C.

ambient temperature and pressure/10% palladium on charcoal for a few hours is a reliable stand-by. But catalyst-poisoning by divalent sulphur (Met, Cys, trace contaminants from the use of sulphur-containing reagents) is a problem. Special procedures, involving the use of liquid ammonia as solvent, or a very large excess of catalyst in acetic acid, have been devised to enable catalytic hydrogenolysis to be performed notwithstanding the presence of sulphur, but these conditions are not well tried. Deprotection by catalytic hydrogenolysis is generally regarded as out of the question if divalent sulphur must be present. This limitation apart, it is a very attractive method.

3.1.2. *t*-Butoxycarbonyl (Boc) protection (6)

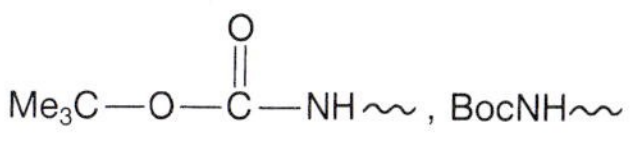

t-Butyl chloroformate, like all alkyl chloroformates which can disintegrate giving a favourable carbonium ion, decomposes very easily (Scheme 3.7), and is unsuitable for preparing Boc derivatives. The

$$Me_3C-O-\overset{\overset{O}{\|}}{C}-NH\sim \; , \; BocNH\sim$$

(6)

$$Me_3C-O-\overset{\overset{O}{\|}}{C}-Cl \longrightarrow Me_3C^+ + CO_2 + Cl^-$$

Scheme 3.7.

corresponding azide, prepared as required from the hydrazide, was for many years the standard reagent, but it has been discredited by explosions, and the anhydride **7** is now preferred. Although rather

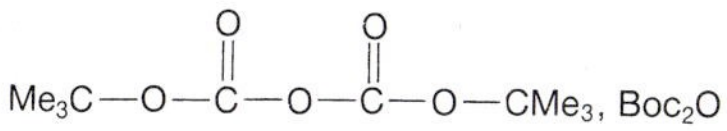

$$Me_3C-O-\overset{\overset{O}{\|}}{C}-O-\overset{\overset{O}{\|}}{C}-O-CMe_3, \; Boc_2O$$

(7)

expensive, it is apparently safe, gives good yields, and it is a convenient solid which can be stored in a refrigerator for long periods without deterioration.

The Boc group is completely stable to catalytic hydrogenolysis conditions (and reducing agents generally), but it is much more labile to acids than the Z group, to which it is therefore orthogonal.

Basic and nucleophilic reagents have no effect at all on the Boc group, even on prolonged exposure. In this respect its stability is better than the Z group's. Hydantoin formation (Scheme 3.3), for example, is not observed with Boc derivatives.

Boc removal is conveniently carried out by dissolution in TFA (either neat or diluted with dichloromethane) at ambient temperature followed after, say, half an hour, by flooding with ether or evaporation. These are mild and reliable conditions. Unimolecular fission, which owes its facility

Two protecting groups are said to be orthogonal if either can be removed without affecting the other (see p. 6).

The indole side-chain of tryptophan (Trp) is electron-rich, and reacts very easily with any electrophilic species generated during acidolytic deprotection: see Section 6.9.

to the stability of the carbonium ion produced, is involved (Scheme 3.8): the *t*-butyl moiety ends up as isobutene unless it is trapped by a

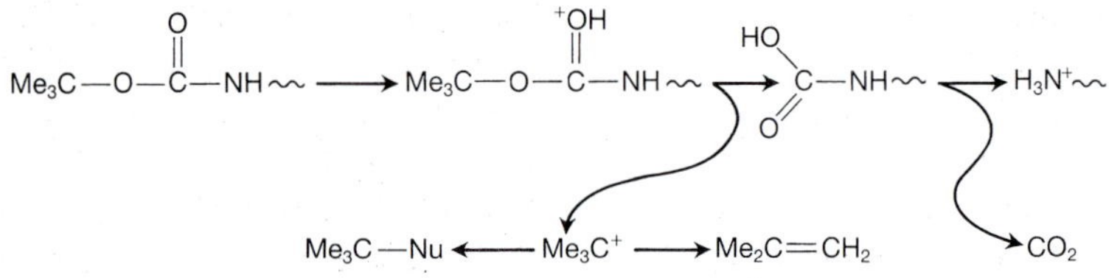

Scheme 3.8. Conditions: TFA.

nucleophile. Scavengers are usually added, and are essential if sensitive residues (e.g. Trp) are present, but they are probably superfluous in simple cases.

Mineral acids attack Boc groups rapidly, and must be avoided in acidification and wash procedures, for which saturated citric acid is commonly used.

Citric acid is a fairly weak acid (first pK_a 3.1), which does not extract easily into organic solvents from water.

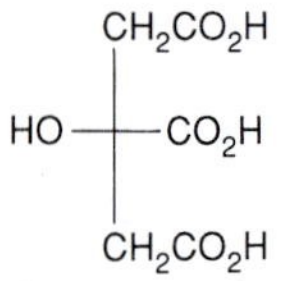

3.1.3   2-(4-Biphenylyl)-isopropoxycarbonyl (Bpoc) protection (8)

The Bpoc group is even more acid-labile than Boc, because the corresponding carbonium ion is not only tertiary but is also further

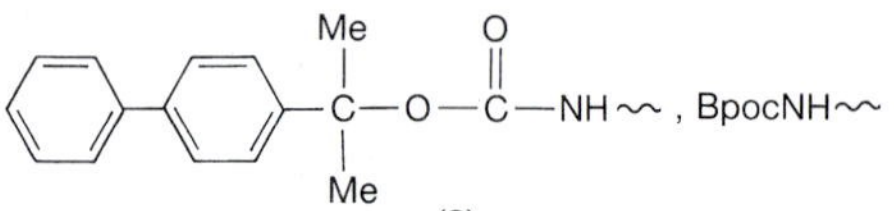

stabilized by the biphenylyl substituent. It can be removed by brief treatment with, for example, chloroacetic acid–dichloromethane mixtures at ambient temperature. These are conditions which leave Boc (and of course Z) groups intact. Bpoc is very stable to bases and nucleophiles, but, like the Z group, it is cleaved by catalytic hydrogenolysis.

3.1.4.   9-Fluorenylmethoxycarbonyl (Fmoc) protection (9)

The Fmoc group is normally introduced in the Schotten-Baumann manner, using the rather stable chloroformate **10**. Dipeptide formation

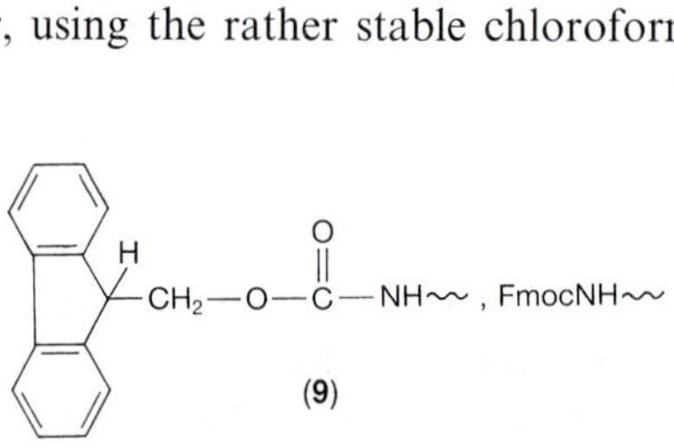

seems to have been found a more serious side-reaction in the preparation of Fmoc amino acids this way than it is with the corresponding reaction of benzyl chloroformate (cf. Scheme 3.2), and tripeptide by-products have been detected too, but this may be due to greater care having been taken to check up on the purity of more expensive intermediates, rather than any chemical difference. Several less reactive reagents have been investigated and recommended: the succinimido ester **11** generally gives results which are superior to those obtained with **10**, and has become quite popular. Alternatively, Fmoc amino acids may be prepared using **10** without oligomer formation by reaction with pertrimethylsilylamino acids and base in aprotic solvents (cf. the preparation of tritylamino acids, Scheme 3.10).

The Fmoc group is very stable to acidic reagents, but is cleaved swiftly under certain basic conditions. Piperidine (20% in DMF) is the routine reagent, but other systems, e.g. fluoride ion in DMF, are also effective. Deprotection with piperidine takes only a matter of seconds at room temperature. The mechanism of cleavage is E1cb, via the stabilized dibenzocyclopentadienide anion; the dibenzofulvene produced reacts with piperidine, giving the adduct **12** as coproduct (Scheme 3.9).

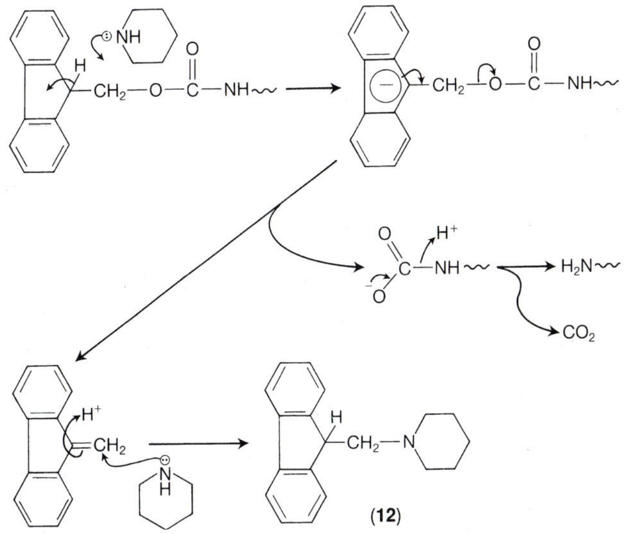

FmocOSu

(11)

The on-off chemistry of the Fmoc analogue Tbfmoc is the same; when it is present it provides a "handle" for purification because it interacts strongly and reversibly with elemental carbon adsorbents.

TbfmocNH ～～～

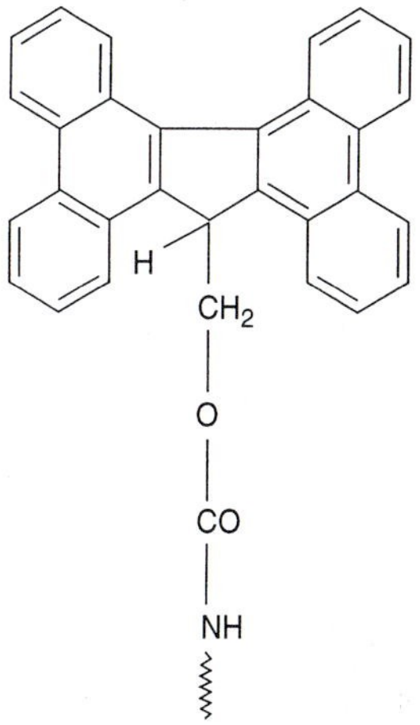

Scheme 3.9. Conditions: 20% piperidine/DMF.

Peptide terminal amino groups, however, neither induce cleavage nor trap dibenzofulvene to a significant extent. This deprotective procedure does not affect Z or Boc groups, or indeed most other modern protecting groups. There was initially some confusion over the position with respect to catalytic hydrogenolysis, but it is now agreed that the Fmoc group is not inert to the usual conditions for carrying this out.

3.2   Other methods

The alkoxycarbonyl family of amino-protecting groups has dominated the field for many years. They will probably continue to do so indefinitely, because the scope they offer for building in relative stability and selective cleavability to order is combined with the fact that the chiral integrity of alkoxycarbonyl amino acids is practically unassailable. This last virtue is not unique to the class, however, and we shall now outline the characteristics of two important protecting groups of very different chemical types which also permit racemization-free manipulations (there are others, too).

Ph$_3$C—NH$\backsim$, TrtNH$\backsim$

(13)

3.2.1.   Triphenylmethyl (trityl, Trt) protection (13)

Yields in the direct tritylation of α-amino acids with trityl chloride under basic conditions are often poor. Reaction with α-amino esters proceeds more satisfactorily, but the subsequent saponification to trityl-amino acids is sluggish, mainly because of steric hindrance, and preparation via pertrimethylsilyl derivatives has been recommended (Scheme 3.10).

$$H_2NCHRCO_2H \xrightarrow{\ i\ } Me_3SiNHCHRCO_2SiMe_3 \xrightarrow{\ ii\ } TrtNHCHRCO_2H$$

(not isolated)

Scheme 3.10. Conditions: i, 2.5 equiv. Me$_3$SiCl/Et$_3$N/CHCl$_3$/reflux/10 min.; ii, 1 equiv. TrtCl/CHCl$_3$/reflux/2 h, then workup after treatment with Et$_3$N/CHCl$_3$-MeOH.

The trityl group is very stable to base, very labile to acid, and does not survive catalytic hydrogenolysis conditions. It predates even the Z group, but has only recently been brought into serious consideration for peptide synthesis, by the trend towards methods employing ever more gentle deprotective reagents: it is so labile to acid, by virtue of the stability of the corresponding cation (Scheme 3.11), that acetic acid or an

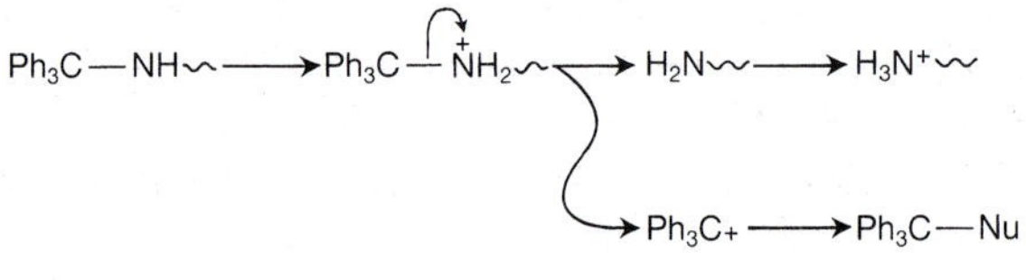

$$Ph_3C - NH\backsim \longrightarrow Ph_3C - \overset{+}{N}H_2\backsim \longrightarrow H_2N\backsim \longrightarrow H_3N^+\backsim$$
$$\searrow Ph_3C_+ \longrightarrow Ph_3C - Nu$$

Scheme 3.11. Conditions: AcOH.

Strictly, the 'pH' scale is only meaningful in dilute aqueous solution, and when peptide chemists refer to 'apparent pH' they usually mean 'apparent pH as judged by the response of water-wet indicator paper to a drop of the test solution'.

apparent pH of 4 in 90% trifluoroethanol (TFE) suffice—conditions which do not affect the Bpoc group. The steric hindrance on which the effectiveness of the group largely depends also influences the accessibility of the carboxy carbon in tritylamino acid derivatives, as the sluggish saponification just mentioned shows. Difficulties in peptide bond formation are therefore to be expected, and most commentators make pessimistic remarks on this point, but the problem has perhaps been made too much of, as some very successful examples of peptide formation with tritylamino acids have been described.

3.2.2 2-Nitrophenylsulphenyl (Nps) protection (14)

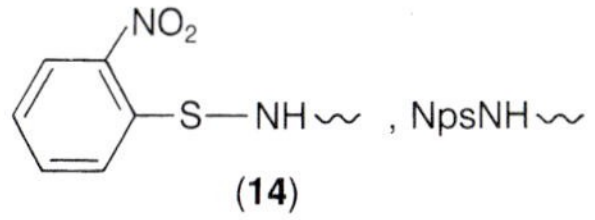

Since Nps amino acids (made by the reaction of NpsCl with amino acids in basic conditions) are not very stable in the free acid state, they are generally isolated, purified, and stored as dicyclohexylammonium salts, from which they can be liberated when required by cautious acidification with sulphuric acid. The Nps group is stable to mild base, but not to acidic conditions or to catalytic hydrogenolysis.

Sulphenamides show evidence of S-N d–p π-bonding, and in Nps derivatives this effect, which would diminish nucleophilic reactivity at the amino nitrogen in any case, is backed up by the electron-withdrawing effect of the 2-nitrophenyl substituent. This nitrogen is therefore well protected, and does not interfere in activation and coupling procedures.

The simplest deprotection procedure is treatment with two equivalents of anhydrous hydrogen chloride in an inert solvent (Scheme 3.12).

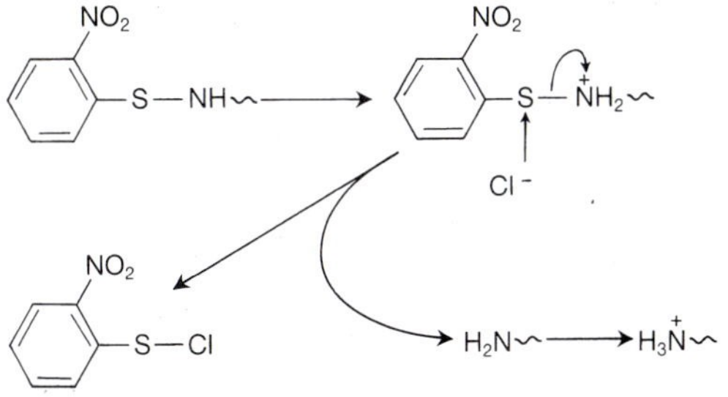

(14)

Scheme 3.12. Conditions: 2 equiv. HCl/Et$_2$O.

This can be performed without cleaving an accompanying Boc group. But side-reactions due to the sulphenyl chloride produced, especially in the presence of tryptophan, are a problem. 2-Mercaptopyridine in acetic acid is selective for Nps cleavage in the presence of t-butanol-derived protecting groups, and does not generate interfering electrophiles.

3.2.3 Dithiasuccinoyl (Dts) protection (15)

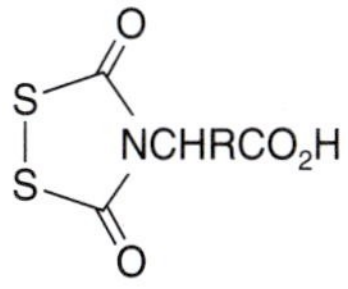

Dts amino acids can be used in standard coupling procedures. They are obtained by successive treatment of amino esters H$_2$NCHRCO$_2$H with (i) EtOCS$_2$Me, (ii) ClSCOCl, and (iii) aqueous acid. The chemistry is complicated and not clean (no doubt it is very smelly too in inexperienced hands!), but this protecting group offers the advantage—not yet fully used—of very selective removability by treatment with a thiol RSH and a mild base such as Et$_3$N. As an exercise, propose plausible mechanisms for the on-off chemistry of Dts protection: see Jones (details in Appendix B, p. 29).

A Dts amino acid

4 α-Carboxy protection

With respect to acylation, the amino groups of amino acids and peptides are considerably more nucleophilic than their carboxy groups are. It follows that amino acids and peptides with unprotected carboxy groups can be used for peptide bond formation, if separately preactivated carboxy components are employed. Many reactions of this kind have been carried out successfully, but a policy of maximal protection is often adopted even when it is not strictly essential, because of solubility advantages, product isolation, and so forth. If they are to be unambiguous, procedures where the carboxy component is activated in the presence of the amino component absolutely require the carboxy group of the latter to be selectively blocked. The usual means of carboxy protection is esterification; the parent carboxylic acid can be regenerated from the esters by acyl–oxygen or alkyl–oxygen fission. The fundamental alkyl–oxygen cleavage chemistry is qualitatively the same as for alkoxycarboxyl groups (see Section 3.1), which are of course esters of a special kind.

4.1 Esters

4.1.1 Methyl and ethyl esters

There is no significant distinction to be drawn between methyl and ethyl esters, and the remarks to be made about the former apply more or less equally to the latter.

Amino acids react easily with hot methanolic hydrogen chloride, to give the corresponding methyl ester hydrochlorides, usually as stable crystalline materials. Treatment with thionyl chloride and methanol is a convenient alternative procedure. The hydrochlorides can be neutralized to give the corresponding free bases, which are distillable liquids in some cases, but which deteriorate quite rapidly at room temperature. The free base forms are therefore generated from the hydrochlorides by neutralization *in situ* when required, normally with a tertiary amino such as triethylamine.

Methyl esters provide good carboxy-protection, and survive well not only most peptide bond forming procedures, but also most deprotective operations. They are not affected at ambient temperature by HBr/AcOH, TFA, catalytic hydrogenolysis conditions, thiols, or amines in organic solvents, so the selective removal of amino-protecting groups from peptide methyl ester derivatives presents no difficulty. The peptide methyl esters are usually isolated as salts, or carried forward immediately. Problems are sometimes encountered at the dipeptide stage, because dipeptide methyl ester free bases cyclize to diketopiperazines (DKPs) rather easily (Scheme 4.1), but methyl esters are otherwise stable and

The decomposition of α-amino esters involves self-aminolysis; diketopiperazines are the main products (cf. Scheme 4.1).

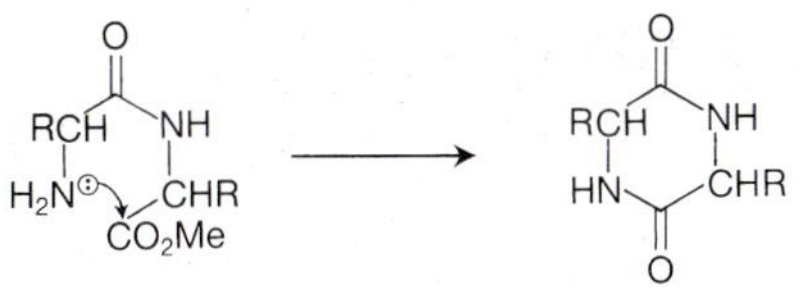

Scheme 4.1. Conditions: basic.

trouble-free protecting groups *en route*. The trouble is that they are a little too stable, and their ultimate deprotection calls for rather vigorous treatment. Saponification is sometimes satisfactory, but even when it is carried out under carefully controlled optimized conditions, alkali may cause racemization and other side-reactions, such as hydantoin formation (see Scheme 3.3). Hydrazinolysis is an attractive alternative when further elaboration of the protected carboxy group is to be carried out, because the hydrazide produced can be used directly in an azide coupling (Section 5.1.1.2), but hydrazinolysis is not an entirely reliable procedure with peptides of any complexity either.

4.1.2 Benzyl esters

Amino acid benzyl esters are best prepared by 4-toluenesulphonic acid-catalysed esterification with benzyl alcohol, driven by azeotropic removal of water in a Dean and Stark apparatus. They are isolated as their 4-toluenesulphonate salts. The corresponding free bases are, like amino acid methyl esters, unstable, and free base dipeptide benzyl esters are prone to DKP formation. Benzyl ester groups are cleaved by saponification and by hydrazinolysis. More importantly, they are cleaved by HBr/AcOH, HF, and by catalytic hydrogenolysis, but not by TFA. Their response to acidic media is similar to that of the Z group, with a slightly lower sensitivity, and the remarks about the mechanism, side-reactions, and substituent effects made in Section 3.1.1 apply here also.

$$H_2NCHRCO_2H/TosOH/BzlOH$$
$$\Delta \downarrow -H_2O$$
$$TosO^-, H_3\overset{+}{N}CHRCO_2Bzl$$

The Z group is after all a benzyl ester too, of a carbonic acid.

4.1.3 *t*-Butyl esters

Amino acid *t*-butyl esters can be prepared directly from the amino acids, but the standard procedure is indirect (Scheme 4.2). Unlike methyl

$$ZNHCHRCO_2H \xrightarrow{\;i\;} ZNHCHRCO_2Bu^t \xrightarrow{\;ii\;} H_2NCHRCO_2Bu^t$$

Scheme 4.2. Conditions: i, CH$_2$Cl$_2$ sat. with Me$_2$C=CH$_2$/H$_2$SO$_4$ (catalytic amount)/20 °C/3 days; ii, H$_2$/Pd(C)/EtOH. Alternatively, stage i may be performed under basic conditions with BocF/Et$_3$N/DMAP/CH$_2$Cl$_2$/Bu^{t}OH/20 °C/4 h.

and benzyl esters, amino acid *t*-butyl esters are stable in the free base form, because the *t*-butyl ester function is very inert to nucleophilic and basic attack, unless there are especially favourable possibilities for intramolecular nucleophilic attack, that is: aspartimides may be formed even from *t*-butyl esters (see Section 6.3), although DKPs are not easily formed from dipeptide *t*-butyl esters.

The stability and lability of *t*-butyl esters generally parallel, as would be expected, the properties of the Boc group. As with benzyl esters versus

See the marginal note above.

Z groups, *t*-butyl esters are slightly less sensitive to acidolysis than Boc groups. The standard TFA treatment cleaves all protecting groups based on *t*-butanol, but, by using carefully controlled exposure to milder conditions, it is in some cases just possible to differentiate between a *t*-butyl ester and a Boc group, and to cleave only the latter. The difference, which is probably to be ascribed to the easier protonation of the carbamate functionality, is not, however, sufficient to be the basis of a routine procedure.

4.1.4   Phenyl esters

Phenyl ester carboxy-protection was devised in response to the need for smooth carboxy-deprotection in the presence of acid-sensitive groups. Phenyl ester salts are easily prepared (e.g. Scheme 4.3), and are usually

$$ZPheOH \xrightarrow{\text{ i }} ZPheOPh \xrightarrow{\text{ ii }} HBr,HPheOPh$$

For DCCI see p. 32.

Scheme 4.3. Conditions: i, PhOH/DCCI/EtOAc/pyridine; ii, HBr/AcOH. Alternatively, the Z group may be removed by catalytic hydrogenolysis: 1 equiv. TosOH/H_2/Pd(C)aq. AcOH.

rather nicely crystalline materials. The group is compatible with standard coupling and isolation routines: it is completely stable to acidic media as well as to catalytic hydrogenolysis conditions. Alkali causes cleavage more rapidly than with methyl esters, but it is the extraordinary susceptibility of phenyl esters to attack by peroxide ion (a so-called *α*-effect nucleophile) which makes this method attractive. Treatment with an equivalent amount of hydrogen peroxide in aqueous DMF, at an apparent pH of 10.5 for 15 minutes, suffices for quantitative cleavage. At this pH, the problems associated with normal saponification are not observed. The cleavage no doubt generates the corresponding peracid first, but peracids are, especially when *α*-electron-withdrawing groups are present, unstable in alkaline solution, so it is the carboxylic acid which is isolated. Peroxide-sensitive residues (Met, Trp, Cys) apparently present no great problem if the deprotection is run in the presence of excess dimethyl sulphide.

It is found that nucleophiles with an electronegative atom having a free electron pair next to the nucleophilic atom show enhanced nucleophilic reactivity versus acyl carbon. This is the '*α*-effect', but it is doubtful whether it can be ascribed to a single factor (polarizability and relatively low steric hindrance are probably significant). Hydrazine also shows enhanced nucleophilic reactivity in acyl transfer reactions.

4.1.5 Phenacyl esters

Carboxy-protection by phenacyl ester ($-CO_2CH_2COPh, -CO_2Pac$) formation is simply achieved by reaction of carboxylate groups with phenacyl bromide, $BrCH_2COPh$. The protecting group has some complications associated with the fact that a ketonic carbonyl group is present, but has the advantage of stability in dry acidic conditions, combined with removability by zinc-acetic acid, making it orthogonal to protecting groups based on *t*-butanol or benzyl alcohol. It is not a universally fashionable protecting group, but is favoured in Japan—for an awe-inspiring example see the final chapter.

5   Peptide bond formation

5.1   The chemical activation and coupling of amino acid derivatives

With very few exceptions, only peptide bond forming procedures involving the generation and aminolysis of reactive carboxy derivatives (Scheme 5.1) have been employed in peptide synthesis so far.

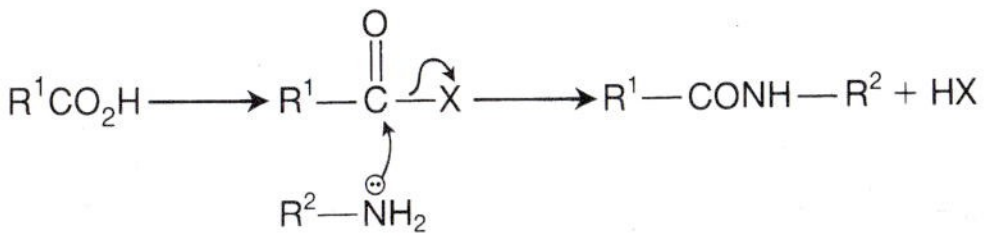

Scheme 5.1.

Activation—i.e. the attachment of a leaving group to the acyl carbon of the carboxy component, to enable attack by the amino group of the amino component—is necessary because ordinary carboxylic acids simply form salts with amines at ambient temperature; the transformation of these salts directly into amides requires severe heating which is quite incompatible with the presence of other functional groups or any structural subtlety. Acidic catalysis, which is effective in bringing about the acylation of simple alcohols by carboxylic acids without the need for other activation, is of no help to direct amide formation because amines are basic, so their availability is diminished under acidic conditions.

There are three different ways of coupling along the lines of Scheme 5.1, although the distinctions are blurred and the fundamental chemistry is universal: (a) those in which a reactive acylating agent is formed from the carboxy component in a separate step or steps, followed by immediate treatment with the amino component; (b) those in which an isolable acylating agent is formed separately, and may even be purified before aminolysis; and (c) those in which the acylating intermediate is generated in the presence of the amino component, by the addition of an activating agent to a mixture of the two components. The activation in (a), (b), and (c) is usually achieved by reaction of the carboxy component with an electrophilic reagent, either by addition (e.g. Scheme 5.2)

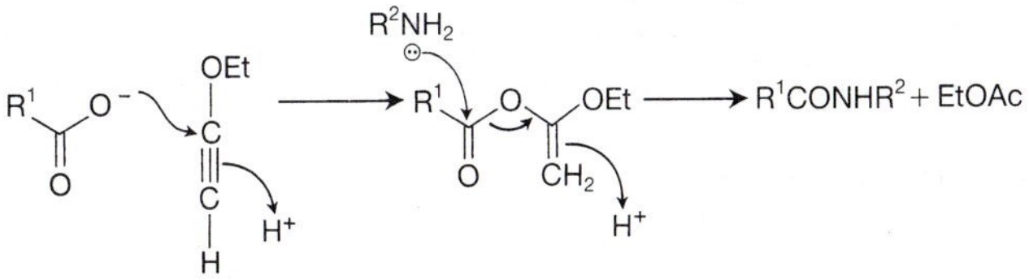

The mechanism of peptide bond formation is actually multistep, involving addition to give a tetrahedral intermediate which collapses in a separate subsequent step. One way of depicting this is as follows:

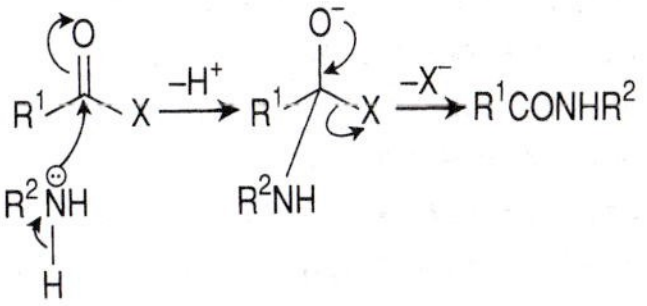

Even this is greatly simplified because the first step is reversible and there is a multiplicity of variations involving proton tranfers to and from O, N, and/or X at different junctures. Some authors conflate the two main steps as follows:

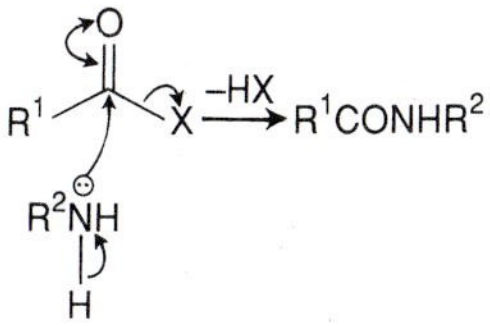

In this book the representation used in Scheme 5.1 will be employed throughout for the sake of simplicity.

Scheme 5.2.   Conditions: R^1CO_2H/R^2NH_2/ethoxyacetylene/moist EtOAc/reflux/1 h.

or by substitution (e.g. Scheme 5.3). It is often the case with type (c) procedures that the activation involves a cascade of acyl-transfer reactions, with more than one species a candidate for consideration as the activated intermediate which suffers aminolysis. Thus in Scheme 5.3,

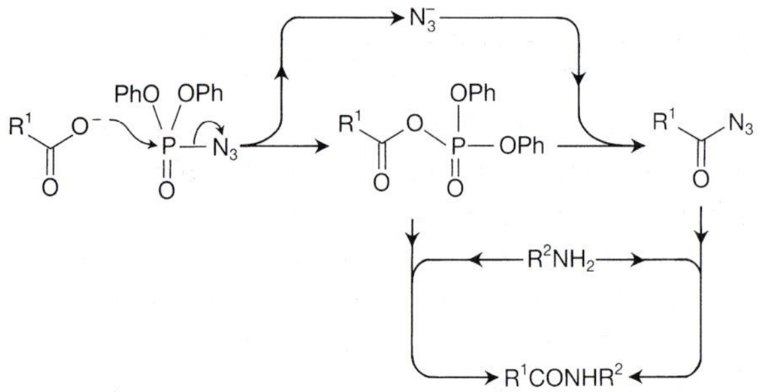

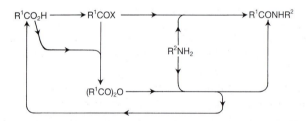

Scheme 5.3. Conditions: R^1CO_2H/R^2NH_2 or $R^2NH_2.HCl/DPPA/Et_3N$ (1 equiv.; 2 equiv. if $R^2NH_2.HCl$ is used)/DMF/0 °C, 2 h to 20 °C, 20 h.

although only two plausible pathways are shown, there is at least one more via the symmetrical anhydride derived from the carboxy component. This might be involved in the ethoxyacetylene reaction (Scheme 5.2) also, and generally to some extent, without making any difference to the overall outcome (Scheme 5.4).

Scheme 5.4.

Ideal chemistry would allow peptide bond formation to be carried out rapidly and quantitatively under mild conditions, avoiding side-reactions, without challenging the integrity of adjacent chiral centres, and generating only easily removed co-products; it would be applicable equally to the addition of one residue at a time ('stepwise synthesis') and the conjunction of long sequences of amino acids ('fragment condensation'). Within the scope of Scheme 5.1, limited though it seems, the imagination of peptide methodologists has found many and diverse possibilities to explore. Scores of these have been shown to be feasible in simple exercises, but a rather small number have been regularly applied in real synthetic problems. It is on these that we shall focus our attention.

5.1.1. Activation and coupling

5.1.1.1 *Acyl chlorides and fluorides*
The activation of acylamino acids by conversion to the corresponding acyl chlorides, followed by reaction with amino acids or esters under

Schotten–Baumann conditions, or by treatment with amino esters in organic solvents, is in principle the most obvious and simple approach to peptide synthesis (e.g. Scheme 5.5), and this kind of method played an

$$\text{PhtGlyOH} \xrightarrow{\text{i}} \text{PhtGlyCl} \xrightarrow{\text{ii}} \text{PhtGlyGlyOH}$$

Scheme 5.5. Conditions: i, PCl_5/PhH/60 °C/30 min.; ii, HGlyOH/MgO/aq. dioxan.

important role in the infancy of the subject. The reagents traditionally used for acyl chloride formation (thionyl chloride, phosphorus pentachloride, etc.) are, however, too vigorous to be compatible with complex or sensitive substrates and, furthermore, most simple acylamino acid chlorides cyclize spontaneously to give oxazolones and hence racemic peptides (see Section 5.1.2.2). Z amino acid chlorides are isolable but unstable, decomposing to *N*-carboxyanhydrides on warming (path a): although they cyclize to give benzyloxyoxazolones under basic conditions (path b), optically active peptides can be isolated after reaction with amino esters, but these and all other kinds of acid chloride were for many years rarely used for peptide bond formation because so many more subtle methods were available. This position was altered dramatically by the discovery that Fmoc amino acid chlorides and also fluorides are easy to prepare, and are rather stable convenient intermediates (e.g. Scheme 5.6). A revised view of acyl halides (especially the fluorides) in peptide

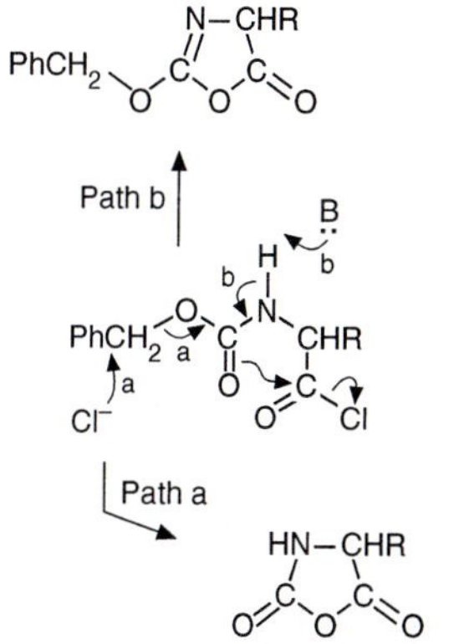

Path b is probably stepwise, with NH ionization ahead of cyclization.

$$\text{FmocValOH} \longrightarrow \text{FmocValCl, m.p. 111–112 °C}$$

Scheme 5.6. Conditions: 10 equiv. $SOCl_2$/CH_2Cl_2/reflux/15 min. Side-chain protecting groups based on *t*-butanol do not survive this procedure because of the HCl produced.

chemistry is being brought about by this, and by the continued development of mild reagents and procedures for their formation.

Fmoc amino acid fluorides can be obtained from Fmoc amino acid pyridine salts by treatment with cyanuric fluoride, a procedure which does not disturb *t*-butyl type protecting groups. Acyl fluorides are more stable than the corresponding acyl chlorides to neutral oxygen nucleophiles such as water, and to tertiary bases, but they show similar high reactivity to amino groups. They are of particular value for forming peptide bonds between sterically hindered amino acids such as α-aminoisobutyric acid (Aib), adjacent residues of which occur in the important antibiotic alamethicin.

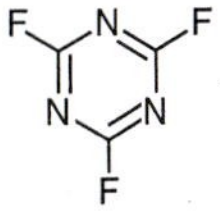

cyanuric fluoride

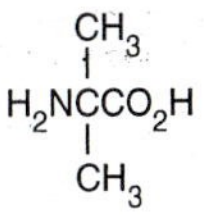

Aib

5.1.1.2 Acyl azides

The use of acyl azides for peptide bond formation was introduced by Curtius a hundred years ago, but is still a significant procedure. The hydrazides from which the azides are generally made (their production from acyl chlorides is of no interest in the present context) may be obtained by hydrazinolysis of a protected amino acid or peptide ester, or by selective deprotection of a peptide derivative which has been constructed using a blocked hydrazide for carboxy-terminal protection. Conversion to the azide was usually achieved in the early days by adding

Azide aminolysis can be sluggish: it is advantageous to run the procedure in the presence of HOCt (see p. 33), which leads to *in situ* formation of a more reactive active ester intermediate.

sodium nitrite to a solution of the hydrazide in a mixture of acetic and aqueous hydrochloric acids, at around or just below 0 °C. This works well in simple cases, but one of the procedures of Honzl and Rudinger (1961), the best of which involves dry acid and a nitrite ester in a dry organic solvent at a lower temperature, is now generally preferred. The azides are subjected to aminolysis without isolation or delay once formed. The essential chemistry of the approach is summarized in Scheme 5.7.

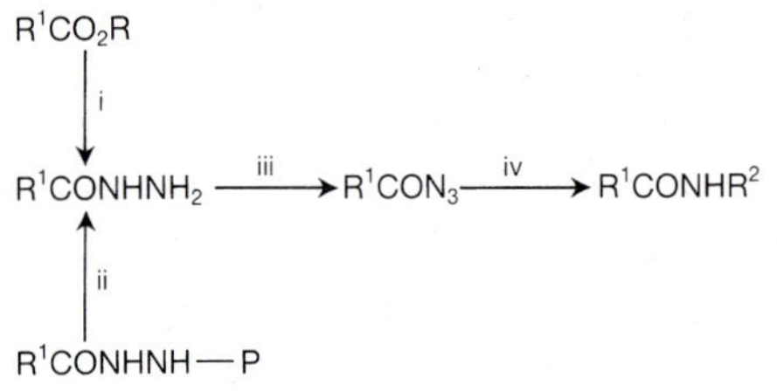

$$R^1CO_2R$$
$$\downarrow \text{i}$$
$$R^1CONHNH_2 \xrightarrow{\text{iii}} R^1CON_3 \xrightarrow{\text{iv}} R^1CONHR^2$$
$$\uparrow \text{ii}$$
$$R^1CONHNH\!-\!P$$

Scheme 5.7. Conditions; i, N_2H_4; ii, deprotection; iii, $NaNO_2$/aq. AcOH/HCl, or an alkyl nitrite ester/H^+ (dry); iv, R^2NH_2.

An obviously possible side-reaction is the Curtius rearrangement of R^1CON_3 to R^1NCO, which might lead to the formation of $R^1NHCONHR^2$.

'Segment' and 'fragment' condensation or 'coupling' are peptide chemists' jargon for procedures in which partial sequences of a target peptide are joined together, as opposed to the stepwise approach in which residues are added one at a time.

Numerous side-reactions attending the formation and use of peptide hydrazides and azides have been recognized, and the azide method has not been used routinely for building up peptide chains stepwise since cleaner methods became available in the 1950s, but it remained completely paramount as a technique for segment condensation until recently. This was partly due to the ways in which the conversions outlined in Scheme 5.7 lend themselves to the production of activated intermediates from protected peptides, but the main consideration was the fact that the coupling of acylpeptide azides was found to be free from racemization. In fact, for many years it appeared that racemization of acylpeptide azides never took place, and confidence in this was one of the cornerstones of peptide synthesis strategy. It is now known that the chiral integrity of acylpeptide azides is not unshakably secure; racemization does occur with excess base in polar solvents, and epimeric peptides have been isolated after fragment condensation by the azide method. Nevertheless, it does seem that the azide technique is less liable to result in racemization than most methods, and with careful control of conditions it can be kept at very low levels.

5.1.1.3   Anhydrides
Symmetrical anhydrides   Acylamino acid symmetrical anhydrides can be prepared (e.g. Scheme 5.8) from the corresponding acylamino acids

$$\text{ZPheOH} \longrightarrow (\text{ZPhe})_2\text{O, m.p. } 139\text{–}140\,°\text{C}$$

Scheme 5.8.   Conditions: 1 equiv. DCCI/MeCN/−5 °C, then 20 °C.

by use of a variety of reagents, including dicyclohexylcarbodiimide (see Section 5.1.1.4). As already mentioned, symmetrical anhydrides may well be the actual acylating species in some procedures involving the direct use of such reagents. Aminolysis of a symmetrical anhydride is unambiguous, but the price for this advantage over some mixed anhydride procedures (see below) is that only half of the carboxy component used is incorporated into the product. The symmetrical anhydrides of Boc, Z, and Fmoc amino acids are generally reasonably stable crystalline substances. That of Z glycine decomposes to **1** on heating, or treatment

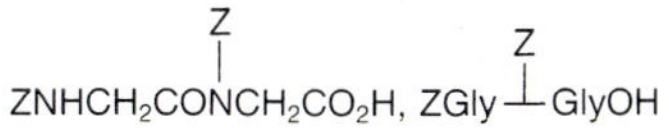

$$ \overset{Z}{\underset{|}{ZNHCH_2CONCH_2CO_2H}}, \quad ZGly\overset{Z}{\underset{|}{\text{—}}}GlyOH $$

(1)

with tertiary base, but otherwise these intermediates appear to be reasonably clean acylating agents, which are very useful in stepwise synthesis. Acylpeptide symmetrical anhydrides would not be appropriate for fragment condensation, even if they could be made without racemization: that would mean a waste of advanced synthetic intermediates.

Mixed anhydrides with carboxylic acids A mixed anhydride of benzoylglycine with benzoic acid was presumably involved in the classic synthesis of benzoylglycylglycine (Scheme 5.9) by Curtius. There are two

Curtius published this work in 1881–1882.

$$ PhCOCl + H_2NCH_2CO_2Ag \longrightarrow PhCONHCH_2CO_2H + PhCONHCH_2CONHCH_2CO_2H $$

Scheme 5.9. Conditions: PhH/heat.

main problems with the use of a mixed anhydride formed between an acylamino acid or acylpeptide and a carboxylic acid: first, the anhydride has two similar electrophilic sites and can therefore undergo aminolysis ambiguously, and second, mixed anhydrides of this sort tend to disproportionate, ultimately with the same result (Scheme 5.10).

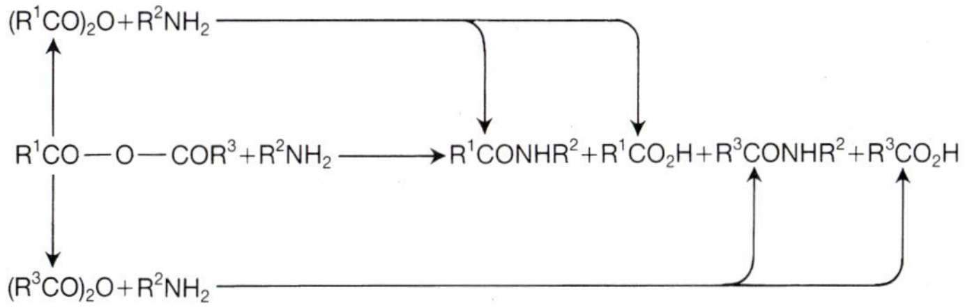

$$ (R^1CO)_2O + R^2NH_2 $$
$$ R^1CO\text{—}O\text{—}COR^3 + R^2NH_2 \longrightarrow R^1CONHR^2 + R^1CO_2H + R^3CONHR^2 + R^3CO_2H $$
$$ (R^3CO)_2O + R^2NH_2 $$

Scheme 5.10.

Pivalic acid is $(CH_3)_3CCO_2H$.

However, the selectivity can be improved greatly by introducing steric hindrance and inductive depression of electrophilicity, so as to direct the attack to the carboxy component carbonyl, as in the case of mixed pivalic anhydrides (e.g. Scheme 5.11), which are the only intermediates in this

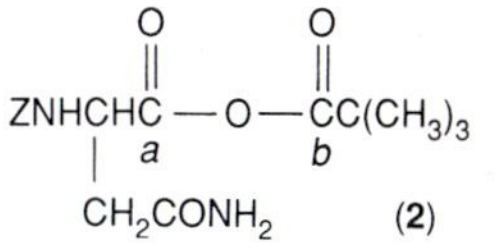

$$ \overset{O}{\overset{||}{ZNHCHC}}\underset{\underset{CH_2CONH_2}{|}}{}\text{—}O\text{—}\overset{O}{\overset{||}{CC(CH_3)_3}} $$
$$ \quad a \qquad\qquad b $$

(2)

$$ ZAsnOH \xrightarrow{i} ZAsnOCOBu^t \xrightarrow{ii} ZAsnCys(Bzl)OMe $$

(2)

Scheme 5.11. Conditions: i, Bu^{t}COCl/N-ethylpiperidine/CHCl$_3$ below $+10\,^\circ$C until solution clear; ii, 5 min. later, HCys(Bzl)OMe/below $+15\,^\circ$C.

In the mixed pivalic anhydride **2** nucleophiles attack preferentially at *a* rather than *b* because that is the least hindered CO, and also because the α-benzyloxycarboxylamino group is slightly electron-withdrawing.

class to have attained significant popularity. Mixed pivalic anhydrides such as **2** appear to be reasonably stable with respect to disproportionation, and can be isolated as crystalline materials in some cases, but they are normally generated and used in separate but rapidly succeeding operations, as in the example of Scheme 5.11. Pivaloyl derivatives of the amino component are occasionally obtained, but this is not normally a serious matter, and the method is a valuable alternative to the mixed carbonic anhydride method, which, for reasons which are not entirely clear, is rather more popular. In fact the two procedures are very similar in the advantages they offer and the problems they present.

Mixed anhydrides with carbonic acids   The most generally successful mixed anhydride method involves the generation and aminolysis of a carboxylic-carbonic anhydride, as outlined in Scheme 5.12. Ethyl

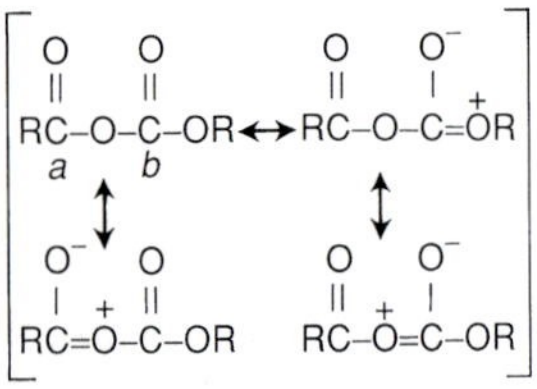

(3)

COa is less deactivated by resonance than COb.

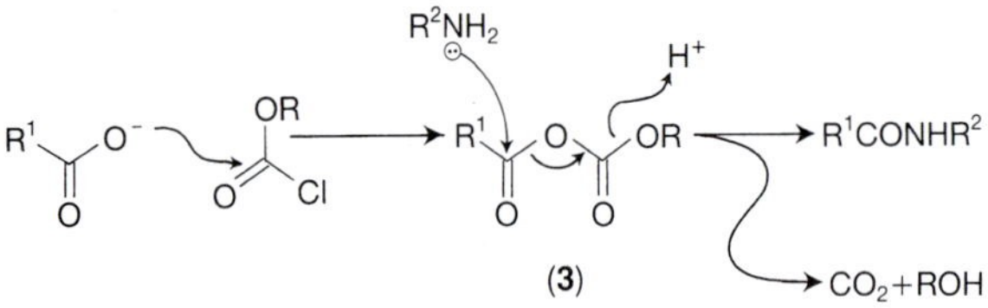

Scheme 5.12. Conditions: i, 1 equiv. tertiary base/unreactive dry solvent/*c.* −10 °C/a few min.; ii, R^2NH_2.

chloroformate is commonly used, but many workers prefer isobutyl chloroformate. The fact that one of the carbonyl groups in the activated intermediate is flanked by two oxygen atoms diminishes its reactivity, so that nucleophilic attack is directed toward the carbonyl of the original carboxy component, and alkoxycarbonyl derivatives of the amino component are not usually formed in more than trace amounts, except with hindered components.

Low temperatures and minimal activation times are usually employed—a typical example is shown in Scheme 5.13—and traditional

$$ZSer(Bu^t)OH \xrightarrow{i,ii} ZSer(Bu^t)GlnGlyOMe$$

Scheme 5.13. Conditions: i, Et_3N/THF/EtOCOCl/−10 °C, followed by ii without delay; ii, HCl.HGlnGlyOMe/Et_3N/DMF added, then −10 °C/1 h, then 20 °C/2 h.

lore stressed the ephemeral nature of the mixed carbonic anhydrides **3**. This now seems rather overdone, because although they do indeed decompose at ambient temperature, it has been shown that alkoxy-carbonylamino acid mixed carbonic anhydrides are in fact isolable. The timescale for their decomposition is measured in hours rather than minutes. Conditions for the minimization of racemization have been delineated for model cases, but the technique has not found wide application for fragment condensation at racemizable residues, as less risky procedures are available. The value of the method lies in its speed and economy, and it has been used in repetitive stepwise synthesis with some success, up to medium chain length. There are some situations where it is especially useful. It often happens, for example, that simple protected peptides in the tri- to hexapeptide range have solubility properties uncomfortably similar to those of dicyclohexylurea, so that the use of dicyclohexylcarbodiimide (see Section 5.1.1.4) is very tiresome; a

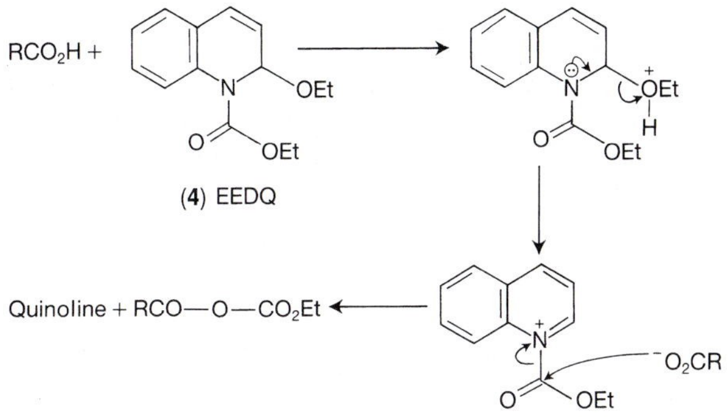

Scheme 5.14.

switch to a mixed carbonic anhydride procedure in such cases may solve the problem.

The classical mixed carbonic anhydride procedure using chloroformates necessarily involves separate activation and aminolysis steps, but mixed ethyl carbonic anhydrides can also be generated by the reaction (Scheme 5.14) of carboxylic acids with 1-ethoxycarbonyl-2-ethoxy-1,2-dihydroquinoline (**4**, EEDQ), and this can be done in the presence of an amino component. Reagent **4** can therefore be used as a direct coupling reagent, as in the example shown in Scheme 5.15. Because in this

$$\text{ZAlaOH} + \text{GlyOEt} \longrightarrow \text{ZAlaGlyOEt}$$

Scheme 5.15. Conditions: 1 equiv. EEDQ/PhH/20 °C/3 h.

technique the mixed carbonic anhydride is consumed by aminolysis as soon as it is formed, the opportunity for the intervention of side-reactions, including racemization when this needs to be considered, is minimal. Since the coproducts are quinoline, ethanol, and carbon dioxide, workup is simple.

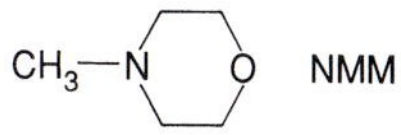

Mixed anhydrides with diphenylphosphinic acid The diphenylphosphinic mixed anhydride procedure which is outlined in Scheme 5.16 is more

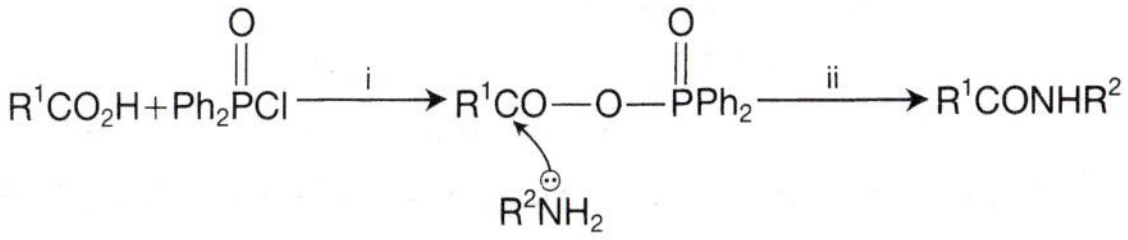

$$\text{CH}_3-\text{N}\bigcirc\text{O} \quad \text{NMM}$$

NMM is the conventional abbreviation for *N*-methyl-morpholine, one of several convenient tertiary amine bases which are used in peptide synthesis. Diisopropylethylamine (DIPEA) is another which is chosen if steric hindrance to nucleophilic involvement is desired; but for the great majority of applications triethylamine is employed.

$$\text{R}^1\text{CO}_2\text{H} + \text{Ph}_2\text{PCl} \xrightarrow{\text{i}} \text{R}^1\text{CO}-\text{O}-\overset{\overset{\text{O}}{\|}}{\text{P}}\text{Ph}_2 \xrightarrow{\text{ii}} \text{R}^1\text{CONHR}^2$$
$$\underset{\text{R}^2\text{NH}_2}{}$$

Scheme 5.16. Conditions: i, NMM/CH$_2$Cl$_2$/−20 °C/20 min.; ii, R^{2}NH$_2$/−20 °C to +20 °C.

regioselective than the mixed carboxylic and carbonic anhydride methods: aminolysis occurs exclusively at the carbonyl of the activated amino acid residue, even when steric factors might militate otherwise. Coupling is thus very clean, and since workup is simple, the procedure is an attractive one.

5.1.1.4   Carbodiimide reagents

Dicyclohexylcarbodiimide (**5**) has been the single most important reagent for activating carboxy groups in peptide synthesis ever since Sheehan and Hess reported their results in 1955, and has only recently started to give way to other reagents. The name of this key reagent is such a mouthful that it is universally called DCC or DCCI, even in speech, and DCCI it will be hereafter in this book.

DCCI may be used to generate activated carboxy derivatives such as symmetrical anhydrides (see Section 5.1.1.3) and active esters (see Section 5.1.1.6) or as a direct coupling reagent. In all cases, the primary activating event is addition of the carboxy group to the carbodiimide functionality to give an *O*-acylisourea (**6**), which is a potent acylating agent (Scheme 5.17). Direct peptide coupling with DCCI is in principle very simple,

Diisopropylcarbodiimide, DICI, and the water-soluble carbodiimide (WSCI) $EtN=C=N(CH_2)_3NMe_2,HCl$, are also used. The chemistry is the same. DICI gives a more organics-soluble urea co-product than DCCI; WSCI gives water-soluble co- and by-products, facilitating purification

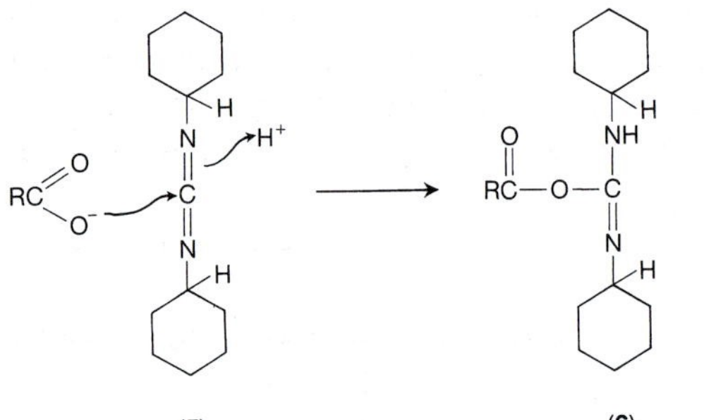

Scheme 5.17.   Conditions: any unreactive solvent; (6) is not isolable.

involving mere mixing of the amino and carboxy components with DCCI in equimolecular amounts in an organic solvent, at ambient temperature or a little below. *O*-Acylisourea formation is rapid, leading to peptide either by immediate aminolysis or via a symmetrical anhydride, with the concomitant formation of dicyclohexylurea (Scheme 5.18). Since the urea

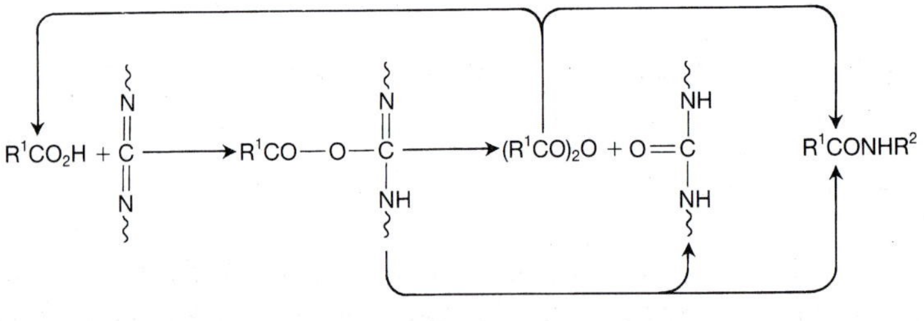

Scheme 5.18.   Conditions: 1:1:1 proportions $R^1CO_2H:R^2NH_2$:DCCI/unreactive solvent/0 °C to 20 °C/a few h.

is only sparingly soluble in most solvents, its separation from the desired product is very straightforward in solution synthesis. It is, on the other hand, soluble enough for removal by thorough washing in solid phase synthesis. But the intermediates shown in Scheme 5.18 are highly reactive, and side reactions can intervene, especially if the amino component is dilatory. Extensive racemization takes place with susceptible carboxy components. Furthermore, the collapse of the *O*-acylisourea by intramolecular acyl transfer sometimes competes significantly with the desired attack by external nucleophiles. When this happens, the much less reactive *N*-acylurea **7** is formed; contrary to what was at one time supposed, this can also be formed by an intermolecular reaction between the urea and a symmetrical anhydride, especially in solvents like DMF. *N*-Acylurea formation not only reduces the yield, but may give rise to purification problems. This difficulty, and the danger of racemization, can both be greatly reduced by performing the coupling in the presence of a suitable *α*-nucleophile which is able to react very rapidly with the *O*-acylisourea before side-reactions can intervene. An acylating agent of lower potency is formed, which is still reactive with respect to aminolysis, but which is more discriminating and does not lead to racemization or other side-reactions. Numerous possible additives, which in effect generate an active ester *in situ*, have been investigated: *N*-hydroxysuccinimide was the first, but 1-hydroxybenzotriazole (**8**,HOBt) has been the one most regularly used so far (Scheme 5.19). The recently introduced aza and monocyclic analogues **9** and **10** may prove to be superior.

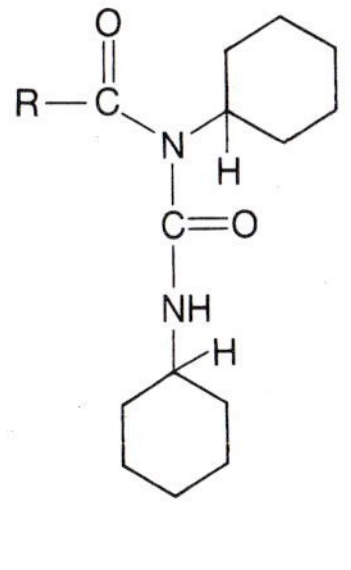

(**7**) *N*-Acylurea

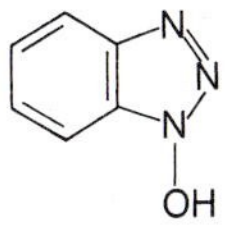

(**8**) HOBt

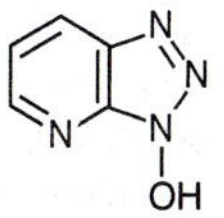

(**9**) HOAt

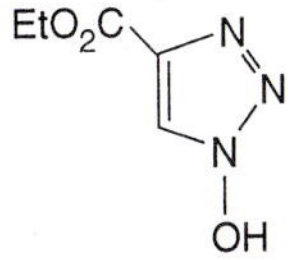

(**10**) HOCt

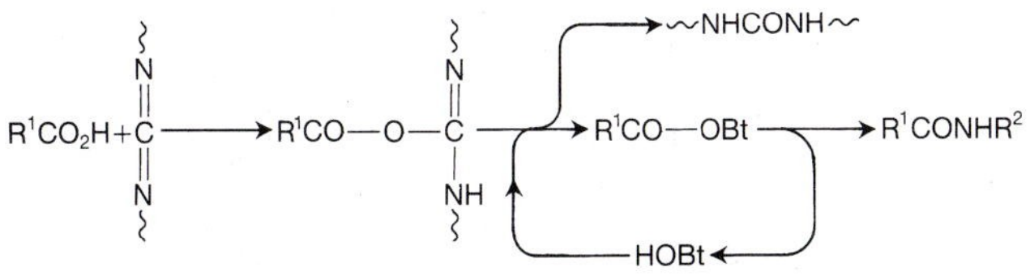

Scheme 5.19. Conditions: 1:1:1:1 proportions $R^1CO_2H:R^2NH_2:DCCl:HOBt/DMF/0\ °C$ to $20\ °C/a$ few h.

5.1.1.5 *Phosphonium and so-called uronium reagents*
Acyloxyphosphonium species, which can be generated by the attack of carboxylate anions on suitable phosphonium cations, react readily with nucleophiles at the acyl carbon (Scheme 5.20), and a number of salts of

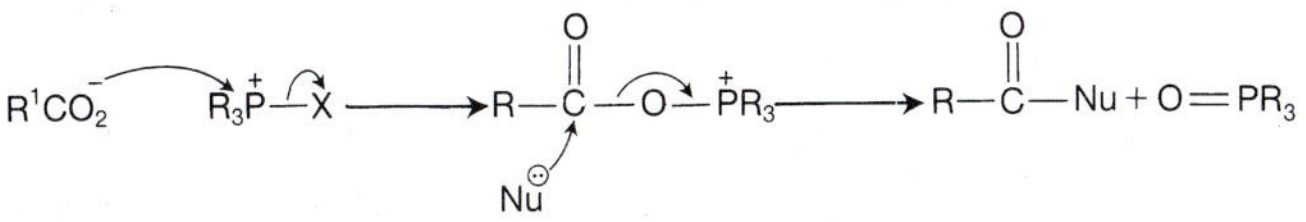

Scheme 5.20.

such cations have been developed for use as direct coupling reagents. Castro's benzotriazolyloxy-tris-(dimethylamino) phosphonium hexafluorophosphate, alias the **BOP** reagent was the trail-blazer. For coupling, an equivalent of the reagent is simply added to a 1:1 mixture,

$(Me_2N)_3\overset{+}{P}{-}OBt,PF_6^-$

BOP

in a suitable inert solvent, of the amino and carboxy components, together with tertiary base to ensure that the latter is in its anionic form. Many pathways are possible, but the main one is via the ester of HOBt (Scheme 5.21). All the co-products are easily removed, and the procedure

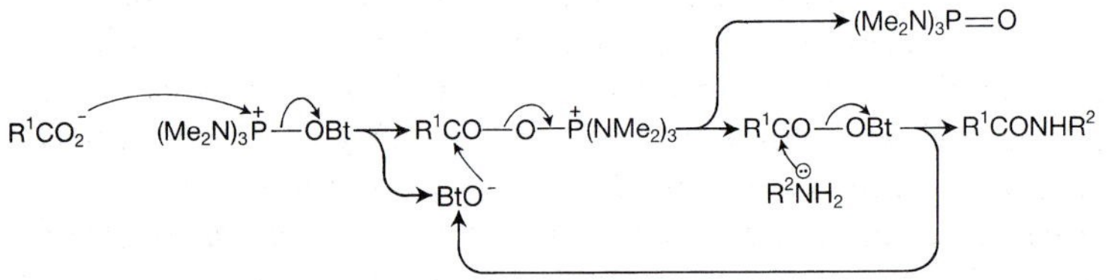

Scheme 5.21. Conditions: 1:1:1:1 proportions $R^1CO_2H/R^2NH_2/BOP/Et_3N$ or, better, DIPEA/ unreactive solvent/20 °C.

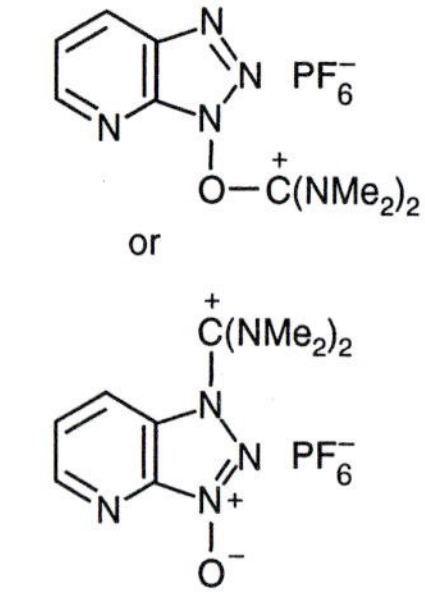

gives excellent yields, with few side-reactions, even in difficult cases. As the initial step is an anion-cation reaction, it is favoured in the less polar solvents, but any inert solvent may be used successfully, including DMF. The BOP reagent is a stable crystalline substance; it is applicable in both solid phase and classical solution work. The first serious threat to DCCI, it advanced rapidly in popularity, even for coupling acylpeptides, despite clear evidence that there was an unacceptable risk of racemization in that application. There was also a safety reservation about the BOP reagent, in that the co-product generated by its use is hexamethylphosphoramide, a highly toxic substance. For this reason numerous closely related phosphonium salts were developed, but these are in turn being pushed aside by reagents such as TBTU and HATU, which work similarly. These reagents are obtained by reaction of **8** or **9** respectively with the tetramethylchloroformadinium cation **11** (generated by the action of phosgene or oxalyl chloride on tetramethylurea). At the outset, TBTU and HATU were formulated and named as uronium salts—hence the U in the acronyms. These acronyms remain in use, despite the fact that it has been discovered that—in the crystalline state at least—both TBTU and HATU are actually guanidinium derivatives. In solution it may be that the two forms coexist: either could react with a carboxyl component to give an activated ester intermediate, so it makes no difference to the outcome. HATU seems to be the best reagent of the class, succeeding in difficult sterically hindered couplings and giving minimal levels of racemization when this is a danger (see Section 5.1.2 below). The use of HATU undoubtedly involves the formation of 7-azabenzotriazol-1-yl esters, which are of very high reactivity to amino groups, probably because of intramolecular general base catalysis as shown in structure **12**.

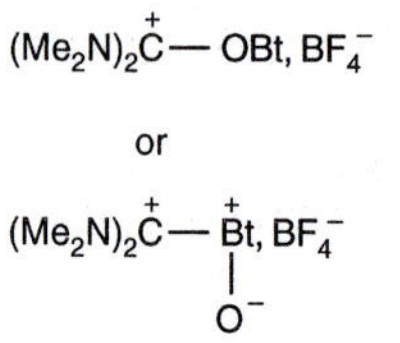

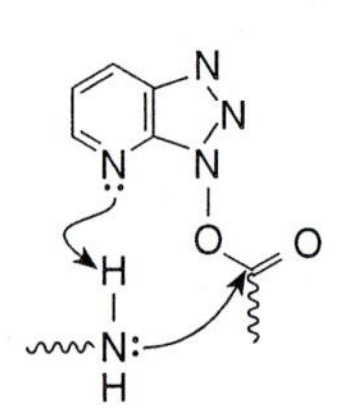

All *in situ* coupling reagents in this diffuse class could in principle, and in all probability do in practice, react by multiple pathways. Two more which are on offer are TFFH and PfPyU (which really must be a uronium salt).

HATU

TBTU

$$(Me_2N)_2\overset{+}{C}\!-\!Cl \quad \textbf{(11)}$$

(12)

As an exercise, see how many plausible pathways you can propose for coupling with these reagents.

$$[Me_2N]_2\overset{+}{C}\!-\!F, \ PF_6^-$$

TFFH

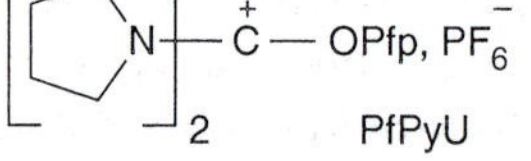

PfPyU

5.1.1.6 Active esters

Attempts to use ester aminolysis (Scheme 5.22) for peptide synthesis were first made in the very infancy of the subject, with little success. It was realized in the early 1950s that the reaction would be facilitated by use of better leaving groups, as in esters of phenols and other similarly acidic functionalities. In contrast to the familiar case of ester saponification, the aminolysis of phenyl esters in organic solvents proceeds with rate limitation by the collapse of the tetrahedral adduct, not its formation, so the leaving ability of the ester group is probably the most important consideration. Since those early days, a very large number of different 'active esters' has been investigated, but only half a dozen or so types (e.g. **13–16**) have been much used. Active esters are most commonly prepared

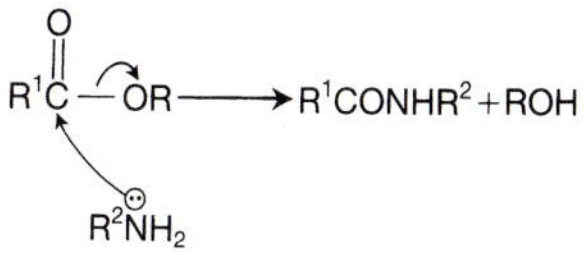

$$R^1C\text{—}OR \longrightarrow R^1CONHR^2 + ROH$$

$$R^2NH_2$$

Scheme 5.22.

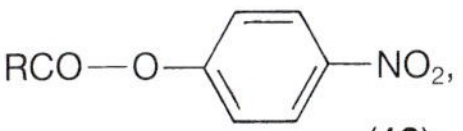

RCO—O—⟨benzene ring⟩—NO$_2$, RCO$_2$Np
(13)

RCO—O—⟨benzene ring with Cl, Cl, Cl⟩—Cl, RCO$_2$Tcp
(14)

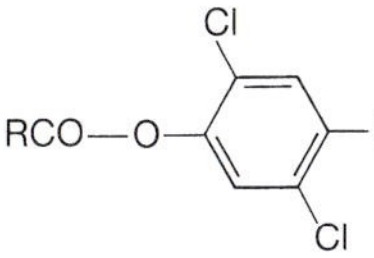

RCO—O—⟨benzene ring with F, F, F, F⟩—F, RCO$_2$Pfp
(15)

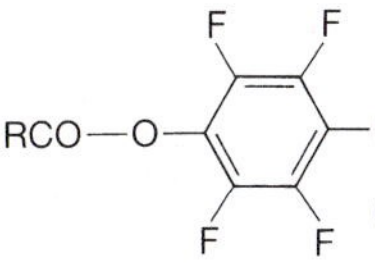

RCO—O—N⟨succinimide ring⟩, RCO$_2$Su
(16)

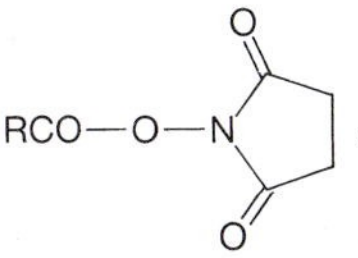

by DCCI-mediated coupling (e.g. Scheme 5.23) between a protected amino acid and the ester moiety. Mixed anhydride methods can also be used, and various specialized reagents are available (e.g. Scheme 5.24). The currently popular active esters are generally crystalline, stable

FmocLeuOH + HOPfp $\longrightarrow$ FmocLeuOPfp

Scheme 5.23.
Conditions: DCCI/dioxan/0 °C.

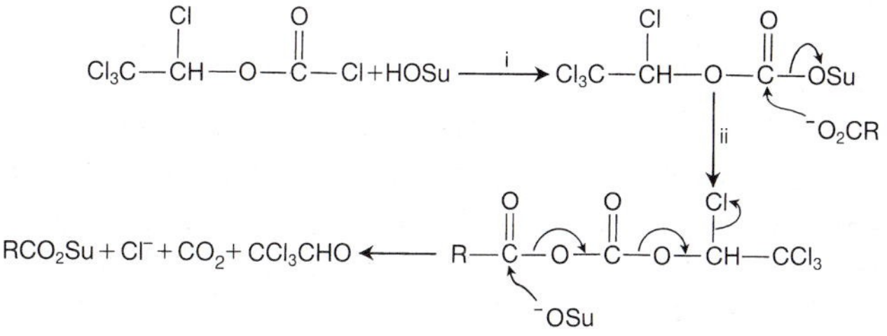

$$Cl_3C\text{—CH—O—C—Cl} + HOSu \xrightarrow{i} Cl_3C\text{—CH—O—C—OSu}$$

$$\downarrow ii \quad {}^-O_2CR$$

$$RCO_2Su + Cl^- + CO_2 + CCl_3CHO \longleftarrow R\text{—C—O—C—O—CH—CCl}_3$$

$${}^-OSu$$

Scheme 5.24. Conditions: i, equiv. amounts of reactants/CH$_2$Cl$_2$/pyridine; ii, RCO$_2$H/NMM/THF/20 °C/2–5 h. The starting chloroformate is freely available (it is prepared by treatment of chloral with phosgene); there are alternative plausible pathways; the chemistry also works for the preparation of other types of active ester.

materials; because they are at a low level of activation (compared to the activated intermediates discussed so far), side-reactions during coupling—including racemization—are generally less of a problem than with most peptide bond forming procedures, although there is a serious risk of racemization when active esters are prepared directly if the carboxy component is susceptible, especially if exposure to base is involved. The cleanliness of the reaction makes the active ester method, especially when the more reactive types of ester are used, appropriate for repetitive classical regimes which dispense with isolation of intermediate peptides, as by-products accumulate slowly. The more reactive active esters are also valuable in solid phase synthesis. Less complete protection than is advisable with more reactive intermediates is often sufficient with an active ester. Although salt couplings, in which the amino component has a free carboxy group, can be performed by other preactivation techniques (but obviously not with direct coupling reagents), active esters give the best results with this device (e.g. Scheme 5.25). Some active ester groups

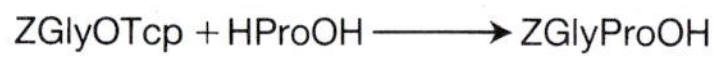

$$\text{ZGlyOTcp} + \text{HProOH} \longrightarrow \text{ZGlyProOH}$$

Scheme 5.25.   Conditions: Et_3N/DMF/2.5 days.

can be used for temporary carboxy blockade (i.e. in 'backing off' operations), which allows indirect access to protected peptide active

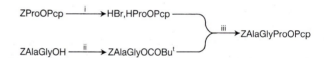

Scheme 5.26.   Conditions: i, HBr/AcOH; ii, 1 equiv. Et_3N, then 1 equiv. $Bu^tCOCl/CHCl_3$/0 °C/ 5 min., then mixed anhydride used without delay; iii, 2 equiv. $Et_3N/CHCl_3$.

esters, e.g. Scheme 5.26; this is sometimes more convenient than their direct preparation from the corresponding carboxylic acids. In such manoeuvres it is clearly essential for acylation to be rapid, or self-condensation of the active ester may occur, so mixed anhydride techniques are preferred.

The reactivities of the various active esters correlate broadly with the acidities of the ester moieties, after due allowance for steric and special effects: **15** and **16** are especially reactive. Choice is partly dictated by sheer reactivity, but the ease of co-product removal is also an important consideration. Thus for a water-insoluble protected peptide product, a succinimido ester coupling is especially convenient, because *N*-hydroxysuccinimide is very water-soluble. But for a water-soluble product, a halophenyl ester may be best, because halophenols are ether-soluble.

5.1.1.7   *N-Carboxyanhydrides*

The *N*-carboxy or Leuchs' anhydrides (NCAs; **17**) of α-amino acids are easily obtainable by the action of phosgene on α-amino acids, or by thermal decomposition of alkoxycarbonyl-α-amino acid chlorides; they are useful for peptide bond formation in two modes.

First, treatment with small amounts of nucleophilic initiator in an organic solvent leads to successive ring opening and loss of carbon dioxide, giving a new nucleophile which attacks another molecule of NCA, and so on, leading to the formation of a homopolyamino acid as outlined in Scheme 5.27, thus making these substances freely available.

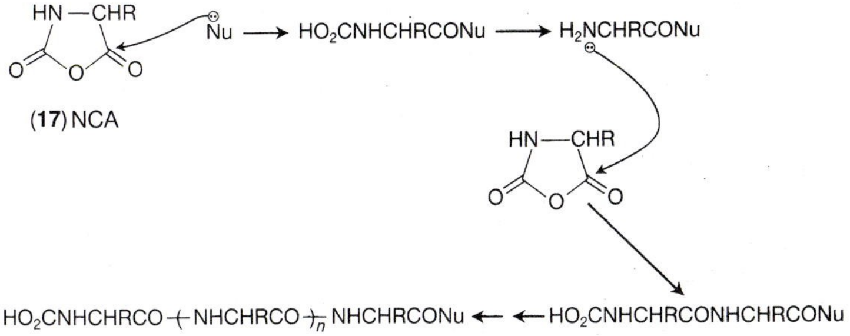

Scheme 5.27. Conditions: a small amount of base or nucleophile/inert solvent/heat. Alternative pathways can operate.

They have been of great service as protein models in diverse biological and physico-chemical studies.

Second, providing the conditions are very carefully controlled, NCAs can be used for rapid peptide synthesis in aqueous solution, as exemplified by the dipeptide synthesis shown in Scheme 5.28. Extension

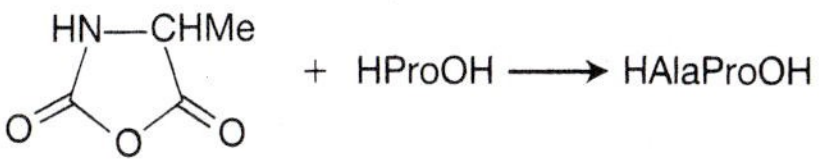

Scheme 5.28. Conditions: NCA added to HProOH/KOH/K_2CO_3/aq. THF/0 °C to +10 °C/15 min., with the pH adjusted to 5.7.

to short oligopeptides is possible by repetition (e.g. Scheme 5.29). The peptides prepared via NCAs in this way are of high optical purity, and in

$$HPheNH_2 \longrightarrow HTrpMetAspPheNH_2$$

Scheme 5.29. Conditions: treatment with the NCA of Asp at pH 10.2, adjustment of the pH to 5 and repetition of the cycle with the NCAs of Met and Trp. Synthesis time 1 h; 30% overall yield of pure material after chromatography.

principle the great rapidity of the technique and the possibility of using a repetitive regime make it very attractive. There is, however, a snag: the carbamic acids produced by aminolysis of NCAs are unstable, and some decarboxylation usually occurs before the aminolysis stage is complete, giving a fresh amino group which can react with any remaining NCA and result in 'over-reaction.' Minimization of this and other side-reactions requires exceedingly careful control of the reaction conditions (and optimization for each case), and the progressive accumulation of over-reaction products limits the extent to which the repetitive approach can be taken without prohibitive purification problems.

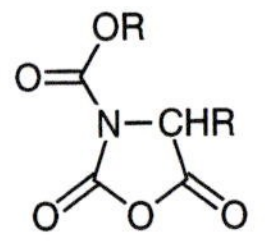

(18)
an UNCA
R=Z, Boc or Fmoc

NCAs can be *N*-alkoxycarbonylated in the presence of base to give urethane-protected NCAs (UNCAs, **18**). The Fmoc-NCAs, which are obtained as shown in Scheme 5.30, are of particular value.

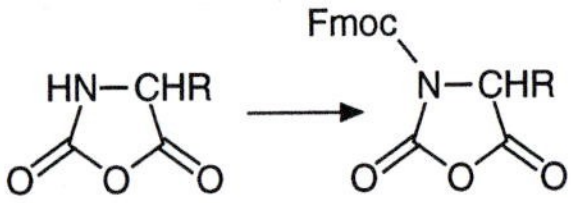

Scheme 5.30. Conditions: dry THF/FmocCl/tertiary amine base.

Fmoc-NCAs are generally nice crystalline compounds, with good shelf-life properties, which react rapidly and cleanly with amino groups to give Fmoc-aminoacyl derivatives and carbon dioxide as the only co-product (Scheme 5.31). Because the product of this aminolysis is safely *N*-protected, over-reaction cannot occur as with simple NCAs: a deprotective protocol must be applied before another residue can be added.

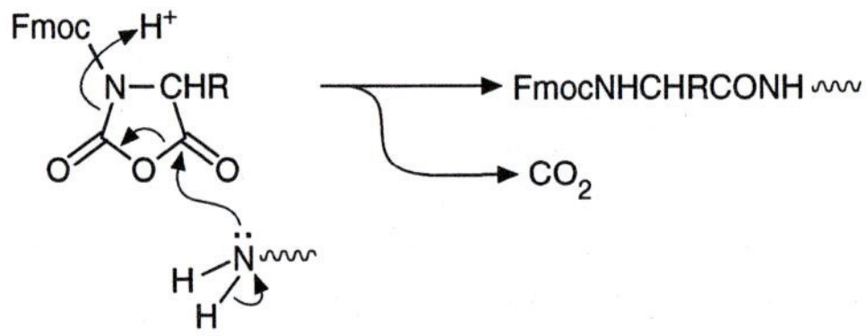

Scheme 5.31.

5.1.2 Racemization

This term is used in peptide chemistry in a loose way which not only covers the strict sense as defined in most general organic chemistry texts (conversion of an enantiomer to a mixture of enantiomers), but also embraces partial epimerization, whereby there is loss of chiral integrity at one of two or more chiral centres, resulting in the formation of a mixture of epimers (i.e. diastereoisomers differing at one chiral centre).

Consider the synthesis of an all-L peptide comprising n chiral residues, from optically pure α-amino acids. If the operations needed for the incorporation of each residue result in conversion of a small fraction f of each residue to the D-form, and the epimers are carried through without separation, then the end product will consist of the required all-L peptide and a blend of other peptides in approximate proportions $(1 - fn)$: fn. For a synthesis of a 50-residue peptide in which 1% D-residue formation takes place at each stage, only half the final product will have the required all-L stereochemistry. The other 50% will consist mainly of about 1% each of all the 50 possible epimers with one D-residue. This will in general pose a prohibitive purification problem, and racemization in peptide synthesis has therefore been closely studied, with a view to defining the conditions under which it is minimal.

Racemization is an almost exclusively base-induced side-reaction, and in practice is only a matter for serious concern at the activation and coupling stages of a synthesis. There are two principal mechanisms.

5.1.2.1 Direct enolization

Deprotonation at the α-carbon of an α-amino-acid residue results in racemization, because the carbanion intermediate can reprotonate on either side (Scheme 5.32). The rate of racemization by direct enolization

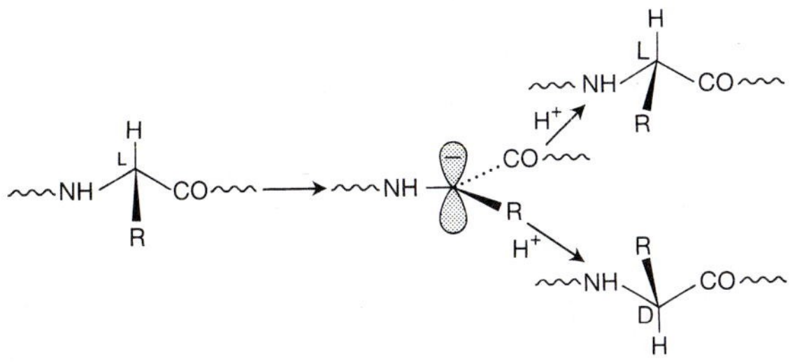

Scheme 5.32

depends on the catalysing base, the solvent, and the electron-withdrawing effects of the groups P,R, and X around the chiral centre (**19**). When X = NH-, O-alkyl, or O$^-$, it is in most cases negligible, and basic deprotection procedures other than saponification are generally completely safe. During activation and coupling, the risk is rather more significant, but the danger is over once coupling is complete. Racemization is fastest with strongly electron-withdrawing groups X (i.e. most good leaving groups), and unhindered strong bases in dipolar aprotic solvents like DMSO and DMF. Inessential exposure to strong bases is clearly to be avoided, but the best way of responding to the other factors is not so obvious, because what matters is not the rate of racemization *per se*, but the balance between the rates of racemization and coupling, and this is much more difficult to make reliable generalizations about. Fortunately, with the exception of (a) a few special amino acids, and (b) couplings which are inordinately slow, the amount of racemization which actually takes place by this pathway is very slight indeed.

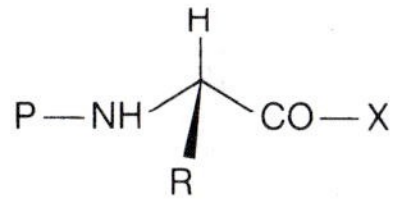

(19)

Certain histidine and cysteine derivatives are somewhat prone to racemization by direct enolization when activated for coupling; α-phenylglycine derivatives are very labile.

5.1.2.2 The oxazolone mechanism

Activated acylamino acids and peptides cyclize under the influence of base to give oxazolones [**20**, strictly '5(4H)-oxazolones', formerly '2-oxazolin-5-ones', archaically 'azlactones']. The oxazolones so formed are themselves activated towards aminolysis, and reaction with amino components leads ultimately to peptides, but since their racemization via stabilized anions is usually fast compared to the rate of peptide bond formation, any peptide thus produced is largely racemized (Scheme 5.33).

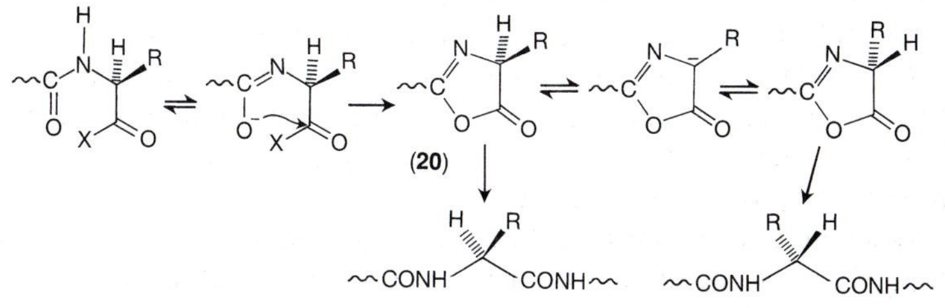

Scheme 5.33. Conditions: basic.

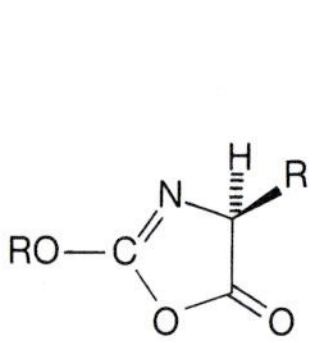

(21)

When the amino-nitrogen of the activated residue is acylated with a simple acyl group (acetyl, benzoyl, etc.), or with a peptide chain, cyclization to the oxazolone occurs easily with most good leaving groups X, and gross or even complete racemization may ensue. But oxazolone formation is not so facile when the acyl substituent is an alkoxycarbonyl protecting group. Indeed, the process was held to be impossible until 1977. Furthermore, the alkoxyoxazolones **21** are both less easily racemized and more easily aminolysed than are the oxazolones **20** derived from simple acylamino acids. The activation of ordinary Z, Boc, and Fmoc amino acids, etc., and their coupling with amino components is consequently not attended by any danger of racemization under normal conditions. This is a pivotal fact on which much of modern peptide synthesis turns. The reason for the contrast between Z and benzoyl amino acids, for example, has not been fully explained, but a major factor is probably the lower NH-acidity of BzlOCONH- compared to PhCONH-.

5.2   The use of enzymes

Nature provides a wide range of proteolytic enzymes which can in principle be perverted to catalyse peptide bond formation instead of hydrolysis by manipulating the conditions. There are two strategies for doing this. The first is dependent on thermodynamic control, the equilibrium in Scheme 5.34 (which favours hydrolysis overwhelmingly

$$R^1CO_2H + H_2NR^2 \rightleftharpoons R^1CONHR^2 + H_2O$$

Scheme 5.34.

under normal conditions) being somehow displaced in favour of peptide bond formation. This can be achieved by employing protecting groups which will ensure precipitation of the peptide, or by using biphasic systems so that the peptide passes out of the aqueous phase into an organic solvent as it is formed, or by using water-miscible organic solvents which perturb the dissociation constants of the components and shift the balance of the equilibrium. The second strategy exerts kinetic control by arranging for an amino component nucleophile to compete with water for an acyl-enzyme intermediate (Scheme 5.35). The

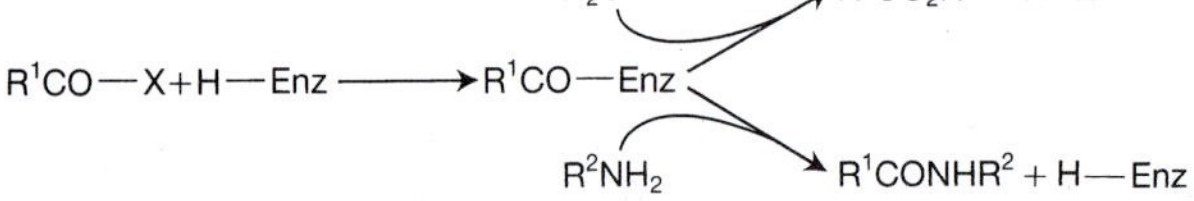

$$R^1CO—X + H—Enz \longrightarrow R^1CO—Enz \xrightarrow[R^2NH_2]{H_2O} \begin{array}{l} R^1CO_2H + H—Enz \\ R^1CONHR^2 + H—Enz \end{array}$$

Scheme 5.35.

advantages of an enzymatic synthesis are the mild conditions, freedom from racemization and the need for side-chain protection, the possibility of using immobilized enzyme technology with catalyst recovery, and the scope for industrial scale-up. Many examples have been reported. The synthesis of the synthetic sweetener aspartame (**22**) is one of particular interest which has been developed for large-scale commercial application, and is even simple enough to be a student exercise (Scheme 5.36). There

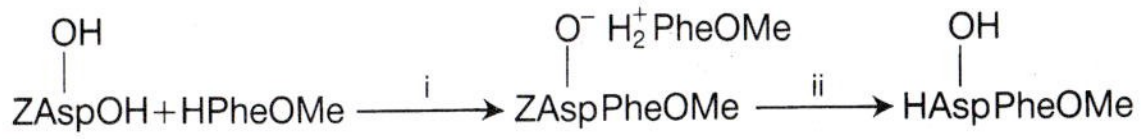

Scheme 5.36. Conditions: i, 2 equiv. HPheOMe/pH7/thermolysin (the dipeptide salt precipitates); ii, HCl, then catalytic transfer hydrogenolysis with HCO_2NH_4/Pd(C)/MeOH.

are disadvantages, however. With peptides longer than dipeptides, there is the danger that while the protease is being persuaded to work backwards in creating a peptide bond at one point, it will remember the purpose for which evolution devised it and dismantle another somewhere else. No new case can be treated as routine. The specificities of the available proteases do not enable all proteinogenic sequences to be assembled, and non-proteinogenic amino acids are not usually acceptable substrates. But steady advances in this area of biotechnology are being made.

5.3   Native chemical ligation

Enzymatic procedures can succeed without vigorous activation of the carboxy component, and do not need protective groups to control selectivity, because the two components are brought together, appropriately oriented for coupling, in an active site which recognises and captures them. This principle can be crudely mimicked with artificial low molecular weight templates, to which the components are separately attached, before peptide bond formation by intramolecular acyl transfer, after which the template is removed. Alternatively, the components can be married through side chains as a prelude to backbone connection.

One such 'native chemical ligation' procedure, devised by Kent and used successfully to couple unprotected peptide components of over 30 residues, is outlined in Scheme 5.37

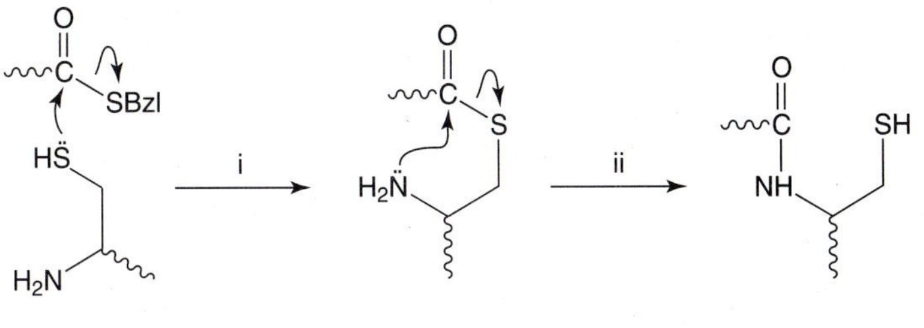

Scheme 5.37. Stages: i, capture of the amino component through its *N*-terminal Cys side chain; ii, intramolecular acyl transfer generating a peptide bond.

The development of procedures for linking large unprotected peptides is a very active area, whether by native chemical ligation through true peptide bonds to give proteins, or by ligation in other ways to give protein-like assemblies. See Lloyd-Williams *et al.* (details in Appendix B), p 175 and reviews by Tam *et al., Biopolymers (Peptide Science),* 1999, **51**, 311 and 2001, **60**, 194.

6 Residue-specific considerations

Every amino acid has unique problems and points of interest: space limits us here to the proteinogenic ones.

6.1 Lysine

Specific peptide bond formation with the α-amino group of lysine requires prior blockade of the ε-amino group. The necessary differentiation can be achieved by using the fact that the side-chain amino group is the more basic and nucleophilic because there is an electron-withdrawing group next to the α-amino group (Scheme 6.1), or by taking advantage

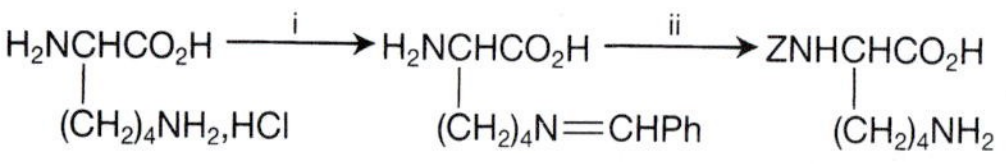

H2NCHCO2H
(CH2)4NH2,HCl
i
H2NCHCO2H
(CH2)4N=CHPh
ii
ZNHCHCO2H
(CH2)4NH2

Scheme 6.1. Conditions: i, 1 equiv. PhCHO/1 equiv. LiOH/H_2O/0 °C to +20 °C; ii, ZCl/NaOH/aq. THF 10 °C/pH 8.8, then brief treatment with warm conc. HCl.

of the fact that the two α-functionalities can engage in chelate formation (Scheme 6.2). Once the amino groups have been distinguished,

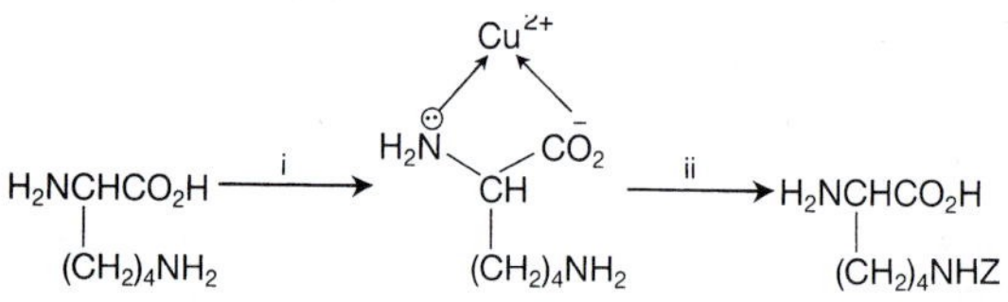

H2NCHCO2H
(CH2)4NH2
i
Cu2+
H2N
CO2
CH
(CH2)4NH2
ii
H2NCHCO2H
(CH2)4NHZ

Scheme 6.2. Conditions: i, aq. $CuSO_4$; ii, ZCl/$NaHCO_3$, then H_2S after acidification.

trivial reactions give ready access to any required orthogonally disubstituted lysines (e.g. the transformations outlined in Scheme 6.3).

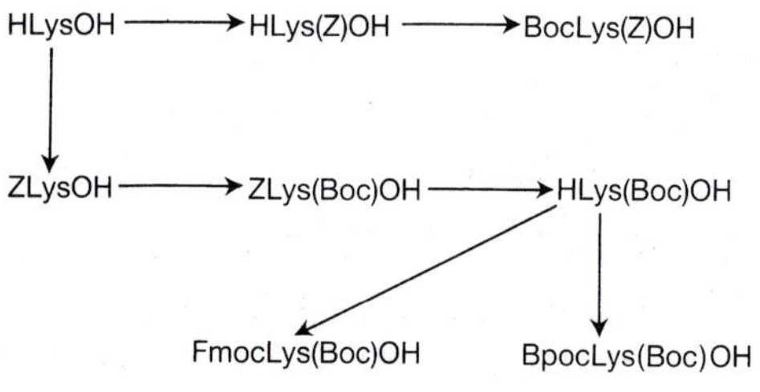

HLysOH
HLys(Z)OH
BocLys(Z)OH
ZLysOH
ZLys(Boc)OH
HLys(Boc)OH
FmocLys(Boc)OH
BpocLys(Boc)OH

Scheme 6.3. Conditions: standard procedures—see Chapters 3 and 4.

The principal ε-protecting groups are Boc and Z (and Z analogues with ring substituents to tune the acid-lability), which have cleavage characteristics much the same as α-Boc and α-Z respectively, although the acid-labilities are slightly greater.

The scope of this book is limited to the synthesis of linear peptides. But cyclic, branched, and conjugated structures involving respectively cyclization, the attachment of other peptides and the conjugation of carbohydrate or other moieties through specific side-chains, are also very important. The construction of such assemblies demands the availability of protective group combinations with additional dimensions of orthogonality. We can illustrate this with three special methods for protecting lysine side-chains:

Alloc $\sim\sim$ NH–CO$_2$CH$_2$CH=CH$_2$

Stable to Boc/OBu^t and Fmoc cleavage conditions, but selectively removed in their presence by Pd^{0}Ph$_4$ catalyzed transfer of the allyl group to mild nucleophiles in almost neutral conditions.

Allyl-based protecting groups are of increasing importance in the synthesis of sensitive polyfunctional compounds.

Mtt $\sim\sim$ NH — C — Ph (with Me-substituted phenyl and Ph)

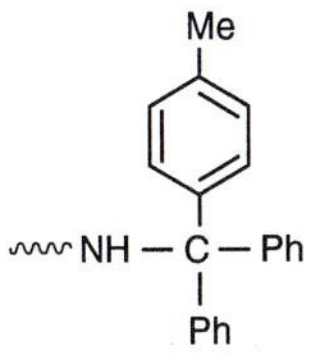

Stable to Fmoc cleavage conditions, but selectively removed by 1% TFA in CH$_2$Cl$_2$, which Fmoc and Boc/OBu^t survive.

Dde

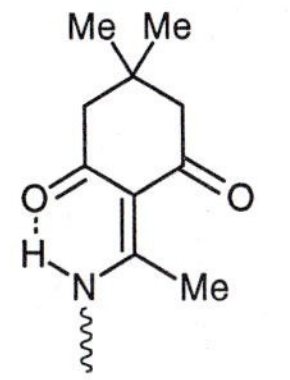

Stable to Fmoc cleavage conditions, but cleaved by 2% NH$_2$NH$_2$ in DMF, which Boc/OBu^t survive.

The cleavage product is

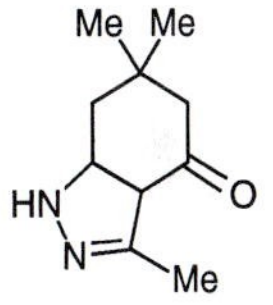

What is the mechanism ?

6.2 Arginine

The guanidino side-chain of arginine is a strong base, which in the unsubstituted state remains protonated under the normal conditions for α-protection, peptide bond formation, and α-deprotection. It is therefore possible to work with unprotected arginine side-chains, but solubility problems and other difficulties may be encountered, and it is generally preferable to protect them. The classic method is nitration (Scheme 6.4):

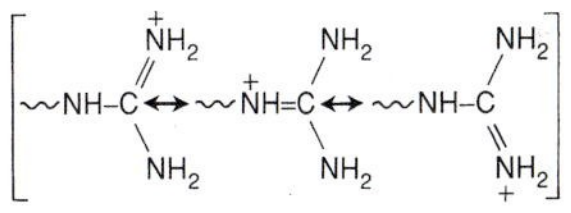

Recall that guanidines are extremely strong bases because of resonance in the corresponding cations.

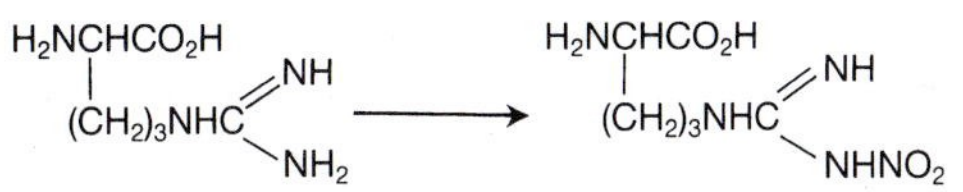

Scheme 6.4. Conditions: fuming H$_2$SO$_4$/fuming HNO$_3$.

ω-nitroarginine has greatly depressed side-chain basicity, and can be α-derivatized for use in peptide synthesis along conventional lines. The ω-nitro group resists HBr/AcOH, but is cleaved to regenerate the guanidino function on treatment with liquid HF, or by reduction with reagents like zinc in acetic acid, or electrolytically, or by catalytic hydrogenation/hydrogenolysis. However, none of these cleavage techniques is entirely satisfactory, and there are also side-reactions in use, notably degradation by hydrazine, and lactam formation in carboxy-activated derivatives (e.g. Scheme 6.5). This side-reaction can also be a

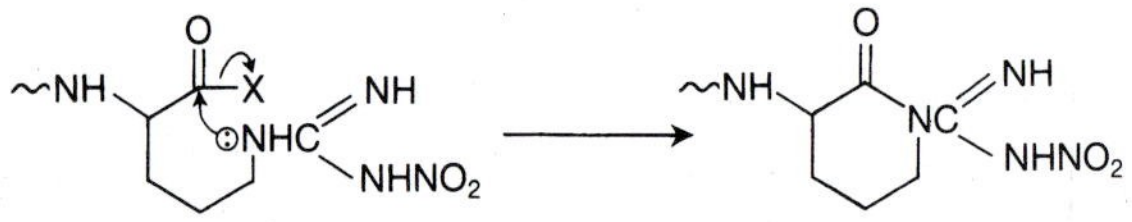

Scheme 6.5. The ring-formation is accelerated by base.

problem with ω-arylsulphonyl protection, which is nevertheless in other respects superior. Tosyl was the first to be tried, and retains a following, despite the fact that cleavage can only be performed with liquid HF or sodium in liquid ammonia: for the preparation of some key intermediates, see Scheme 6.6. Acid-sensitivity is enhanced by the intro-

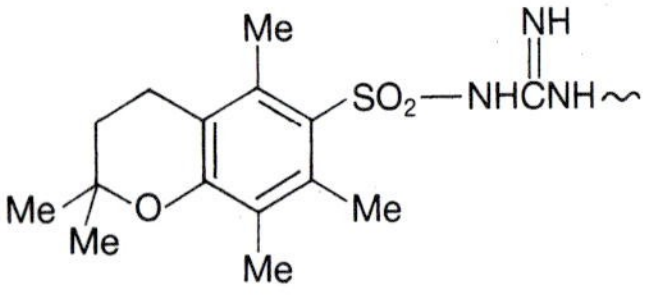

(1) Pmc

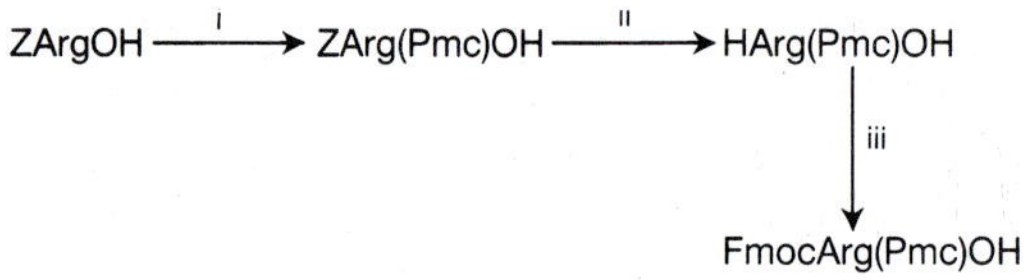

$$ZArgOH \xrightarrow{\ i\ } ZArg(Pmc)OH \xrightarrow{\ ii\ } HArg(Pmc)OH$$
$$\downarrow iii$$
$$FmocArg(Pmc)OH$$

Scheme 6.6. Conditions: i, PmcCl/NaOH/aq. Me$_2$CO; ii, H$_2$/Pd(C)/MeOH; iii, FmocOSu/Na$_2$CO$_3$/aq. DMF.

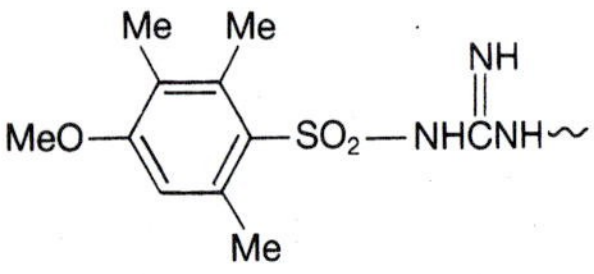

duction of electron-releasing groups, provided counterproductive stereoelectronic effects are not simultaneously imposed. The limit appears to have been approached with the 2,2,5,7,8-pentamethylchroman-6-sulphonyl (Pmc) group (**1**), which is cleaved by TFA under conditions comparable to those required for Boc cleavage: FmocArg(Pmc)OH (Scheme 6.6) is the key intermediate here. The 4-methoxy-2,3,6-trimethylbenzenesulphonyl (Mtr: **2**) group is somewhat less acid-labile, requiring a few hours in TFA for cleavage.

(2) Mtr

6.3   Aspartic and glutamic acids

Controlled peptide bond formation involving the α-carboxy group of either aspartic or glutamic acid absolutely requires obstruction of the other carboxy group. Benzyl or *t*-butyl ester protection is usually employed. The necessary intermediates can be obtained by, *inter alia*, the routes outlined in Scheme 6.7 (illustrated for Glu: analogously for Asp). The differentiations between the carboxy groups shown depend, in the case of the acid-catalysed esterification, on the fact that the α-carboxy group is deactivated by a nearby protonated amino group, and, in the case of the anhydride ring-opening, on the fact that the more electrophilic carbonyl is the one adjacent to the electron-withdrawing acylamino function. The cleavage characteristics of side-chain esters are essentially the same as α-esters.

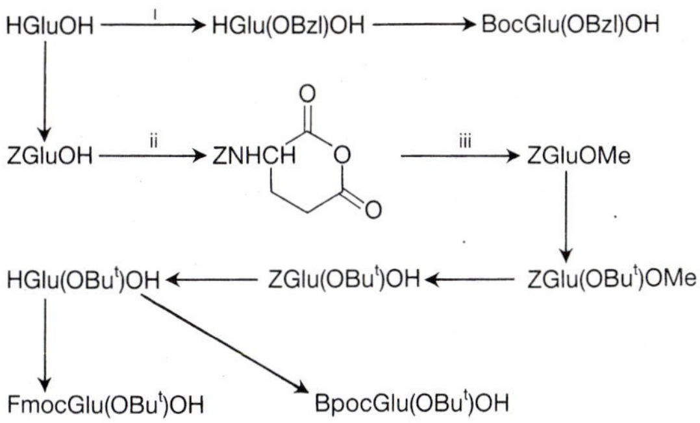

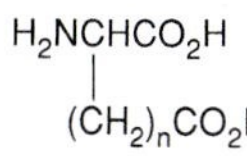

Scheme 6.7. Conditions: i, BzlOH/H$_2$SO$_4$/20 °C followed by neutralization (single isomer, 75% yield, crystalline); ii, Ac$_2$O/20 °C until solution is complete, followed by evaporation at 50 °C; iii, MeOH/dicyclohexylamine/Et$_2$O/20 °C/12 h, followed by collection of the salt (a single crystalline isomer, 55% yield) and neutralization. The rest of the conversions may be achieved by standard procedures (see Chapters 3 and 4).

The main difficulties encountered with aspartic and glutamic acids stem from the ease with which cyclic imide and amide rings form. Thus aspartylpeptides, especially aspartylglycine- and aspartylserine-containing sequences, are liable to cyclize to five-membered imides, under acidic or basic conditions or both, depending on the kind of ester if any (Scheme 6.8). Indeed, this is perhaps the most serious side-reaction encountered in

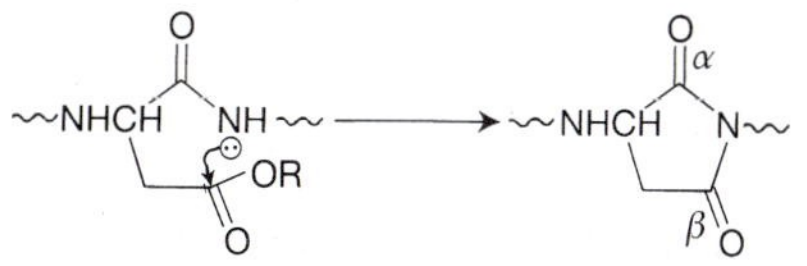

Scheme 6.8. Hydrolytic ring-opening at the β-carbonyl regenerates the original peptide (less the side-chain ester), but attack at the α-carbonyl, which is favoured, gives a β-linked peptide.

peptide synthesis, and is particularly insidious because routine process control by amino-acid analysis cannot reveal it. It is less serious with secondary and tertiary alkyl esters in the side-chain than it is with esters of primary alcohols. Imide formation is not so facile when a six-membered ring is concerned, so it is less of a problem with glutamic acid derivatives—but for these five-membered cyclic amide formation is a nuisance under some circumstances (see Section 6.7).

For this reason cyclohexyl (cHx) esters are increasingly favoured over the classical benzyl esters for side chain protection in conjunction with α-Boc: stable to TFA, cleaved by HF.

6.4 Cysteine and cystine

Cysteine presents a unique synthetic problem. Its side-chain is more than a functional appendage. In most cases it forms, through disulphide formation between two cysteine residues to give a cystine residue, part of the molecular framework of the peptide it belongs to.

Until the time is ripe in a synthesis for disulphide formation, the cysteine thiol groups must be protected, because they interfere with many routine operations, but their soft nucleophilic reactivity does at least facilitate selective S-protection by direct alkylation of cysteine itself: see

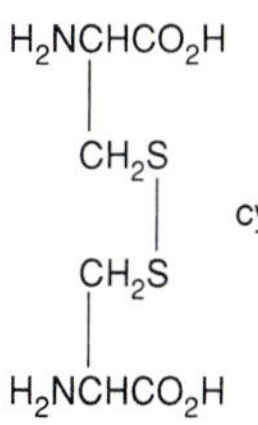

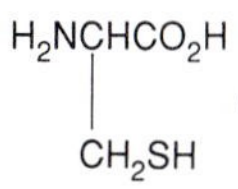

Scheme 6.9. *S*-Acylation is of little use as a basis for protection, because

S-Protection cleaved by I_2
with S—S formation;
also cleaved by strongly
acidic media
HCys(Trt)OH
ii
HCys(Bzl)OH
S-Protection cleaved by
$Na/NH_3(l)$ or, rather
slowly, HF
i
HCysOH
iii
HCys[Bzl(OMe)]OH
S-Protection cleaved
by HF, faster
iv
HCys(Acm)OH
S-Protection cleaved by I_2
with S—S formation;
survives HF

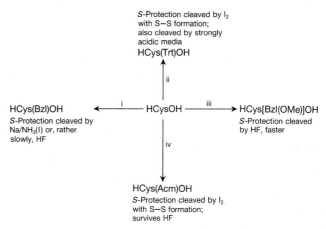

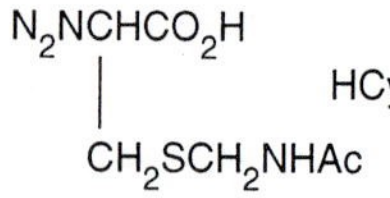

N_2NCHCO_2H
CH_2SCH_2NHAc

HCys(Acm)OH

Scheme 6.9. Conditions: i, BzlBr/aq. NaOH/5°C; ii, TrtOH/AcOH/BF_3Et_2O/60°C; iii, Bzl(OMe)Cl, cf. i; iv, $AcNHCH_2OH$ (i.e. AcmOH)/TFA/20 °C/0.5 h. All except Cys(Bzl) may also be cleaved, in some cases with direct disulphide formation, by reagents such as sulphenyl chlorides, $Hg(OAc)_2$, and $Tl(CF_3CO_2)_3$.

thiol esters are too reactive towards nucleophiles. Apart from the thioether-type groups of graded cleavability, some of which are shown in Scheme 6.9, the only other significant approach is mixed disulphide formation with *t*-butyl mercaptan (Scheme 6.10).

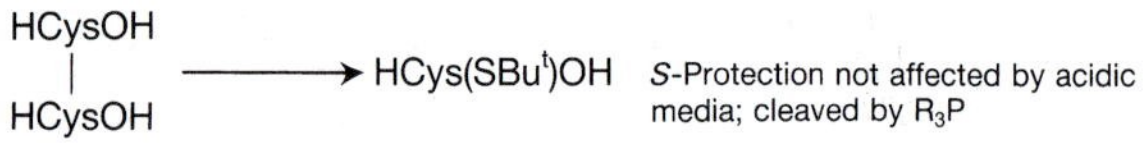

HCysOH
HCysOH
HCys(SBu^t)OH
S-Protection not affected by acidic
media; cleaved by R_3P

Scheme 6.10. Conditions: NaOH/aq. dioxan/Bu^{t}SH/air stream/20 °C.

None of the *S*-protecting groups indicated is compatible with ordinary catalytic hydrogenolysis, but H_2/Pd in liquid ammonia may be used for removing Z from Cys(Bzl) derivatives. Mild acid and base α-deprotective procedures may be used in conjunction with them.

Peptides with a single intrachain disulphide bridge, and symmetrical cystine peptides, may be made simply by *S*-deprotection and then oxidation (Scheme 6.11), as exemplified by the synthesis of oxytocin

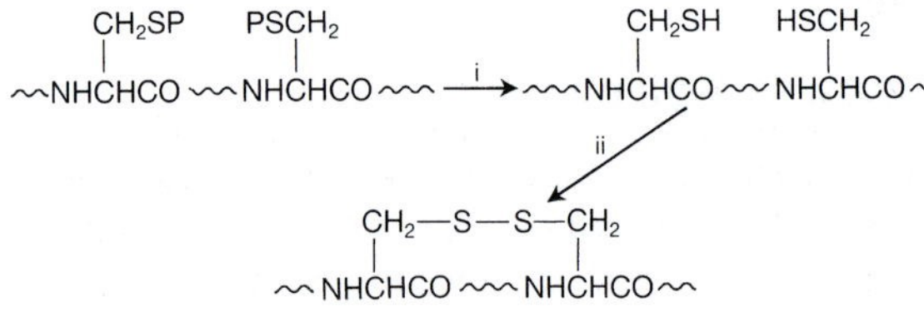

CH_2SP
$PSCH_2$
NHCHCO
NHCHCO
i
CH_2SH
$HSCH_2$
NHCHCO
NHCHCO
ii
CH_2—S—S—CH_2
NHCHCO
NHCHCO

Scheme 6.11. Conditions: i, deprotection; ii, oxidation.

(see Section 8.2.1). Alternatively, reactions are available for bridging in one operation without separate deprotection. The iodine oxidation of bis-trityl peptides (Scheme 6.12) was the first procedure of this class.

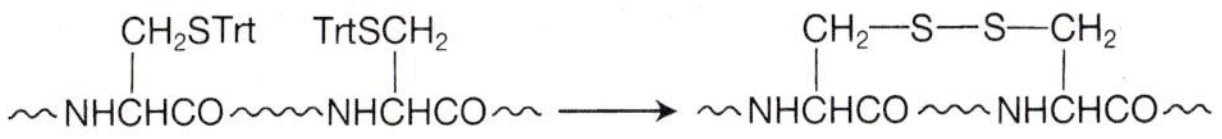

Scheme 6.12.   Conditions: I$_2$/protic or aprotic solvent.

The reaction, which is applicable to other *S*-protecting groups, notably Acm, probably proceeds as outlined in Scheme 6.13. The rate is very

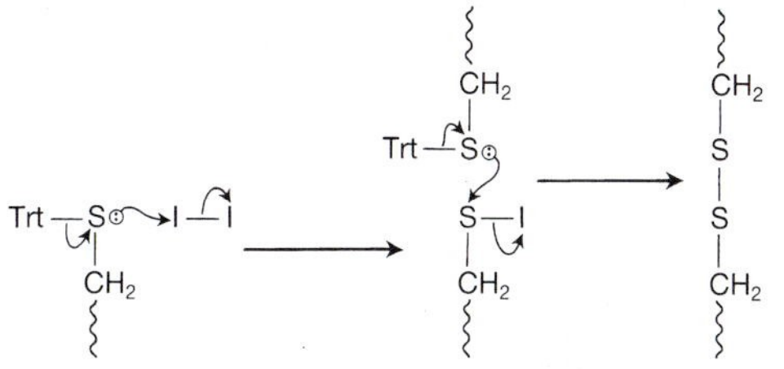

Scheme 6.13.

dependent on the protecting group and on the solvent, and appropriate choice of conditions allows selectivity when both Cys(Trt) and Cys(Acm) are present. Peptides with two or more disulphide bridges in different regions of a single chain can be approached via fragments with preformed bridges, but more complicated topology necessitates controlled formation of the disulphide links in turn. For this, sets of *S*-protecting groups of differentiated reactivity are required, and for unsymmetrical interchain links it has also been necessary to devise means of driving the bond formation in the desired direction at the expense of symmetrical dimer formation (e.g. Scheme 6.14). These principles are illustrated in the syntheses of insulin (Section 8.2.3).

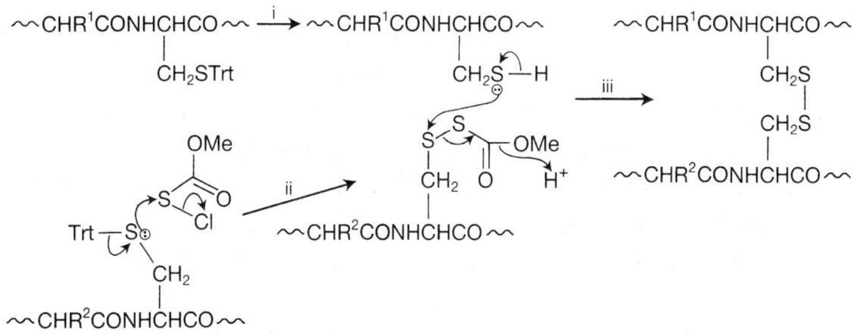

For a full account of disulphide bridge formation see the chapter on the subject in Lloyd-Williams *et al.* (details in Appendix B).

Scheme 6.14.   Conditions: i, Hg(OAc)$_2$/EtOAc-MeOH, then H$_2$S; ii, MeOC(=O)SCl (i.e. ScmCl)/CHCl$_3$-MeOH/0 °C/1 h; iii, equivalent amounts of the products from i and ii/CH$_2$Cl$_2$/1.5 h/20 °C. The conversion i proceeds equally well with the corresponding Cys(Acm) derivative.

6.5   Methionine

Methionine inhibits freedom of synthetic manoeuvre, because the thioether side-chain interferes with catalytic hydrogenation, and is liable to be attacked by electrophiles generated on acidolytic deprotection of other functionalities. The usual response to its presence is to live with it, adjusting the conditions chosen accordingly, but protection by reversible oxidation to the sulphoxide (Scheme 6.15) is favoured by some workers.

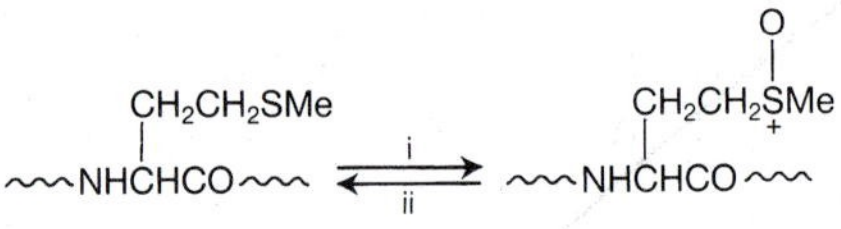

Scheme 6.15. Conditions: i, H_2O_2 (slightly more than 1 equiv.)/EtOH/23 °C/4 h; ii, H_2O/excess $HSCH_2CO_2H$/under N_2/50 °C/20 h.

6.6   Serine, threonine, and tyrosine

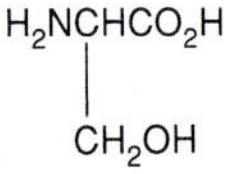

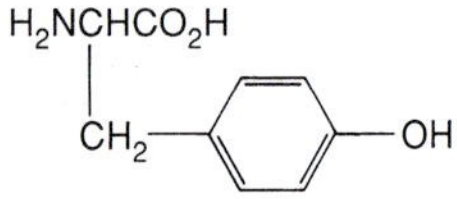

Hydroxy groups, especially phenolic hydroxy groups, react with acylating agents, and are therefore usually protected in peptide synthesis, but they are sometimes left unblocked in syntheses involving only mild carboxy-activation. Benzyl ether protection, cleaved by vigorous acidolysis (HF,HBr/AcOH, etc) or by catalytic hydrogenolysis, or *t*-butyl ether protection, cleaved by TFA, is standard: all the requisite doubly protected intermediates are accessible, albeit in some cases by round-about routes (Schemes 6.16 and 6.17 illustrate some key reactions for Thr

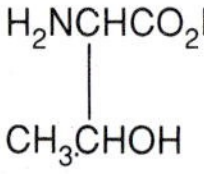

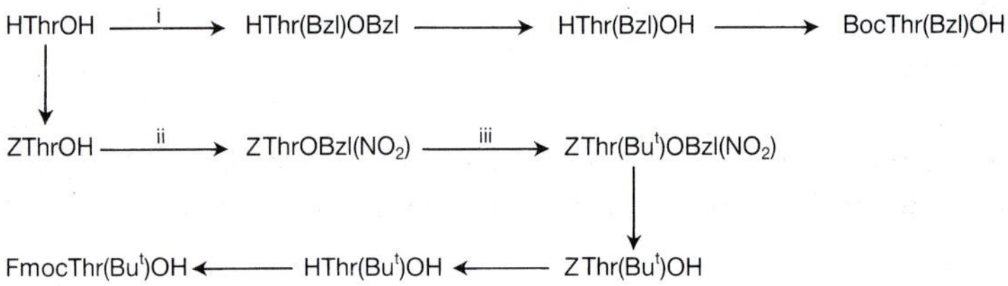

Scheme 6.16. Conditions: i, BzlOH/PhMe/TosOH/heat until no more water is produced (product isolated as its oxalate, pure but in low yield); ii, Et_3N/1 equiv. $Bzl(NO_2)Br$/EtOAc/reflux/ 8 h; iii, $Me_2C=CH_2$/H_2SO_4/CH_2Cl_2/20 °C/4 days. The other conversions may be achieved by essentially standard procedures (see Chapters 3 and 4).

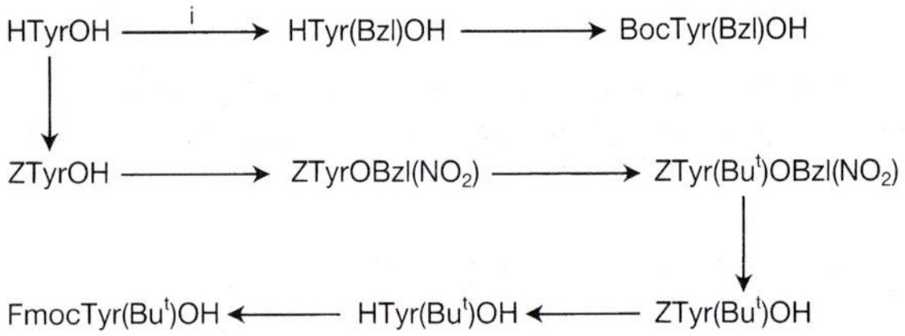

Scheme 6.17. Conditions: i, $CuSO_4$/aq. NaOH, then BzlBr/25 °C/1 h, then acidification to destroy the copper complex. The other conversions may be achieved by essentially standard procedures (see Scheme 6.16 and Chapters 3 and 4).

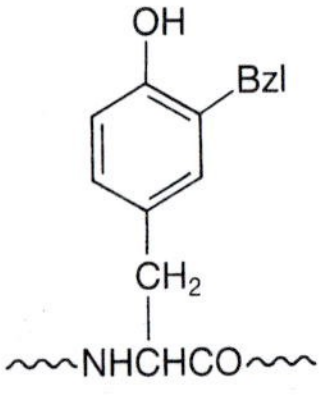

and Tyr). The acid-lability of benzyl and *t*-butyl ethers is comparable to that of the other protecting groups based on benzyl and *t*-butyl alcohols. The only major problem arises from the susceptibility of the phenolic ring to attack by electrophiles. Cleavage of *O*-benzyltyrosine protection with HF or HBr/TFA leads to the formation of 3-benzyltyrosine residues **3**, a side reaction which can be diminished by shifting the mechanism in the more S_N2-like direction. This can be done, by using HBr/AcOH rather than HBr/TFA; by the addition of soft nucleophiles like thioanisole; and by placing electron-withdrawing substituents on the ring of the protecting group. Alternatively, 2-bromobenzyloxycarbonyl protection (**4**) is usually stable enough to nucleophiles (despite its phenyl ester character) and to TFA for use in Merrifield solid phase work, but is cleanly cleaved by strong acidolysis along with other conventional side-chain protection, and this is now the favoured choice in that field.

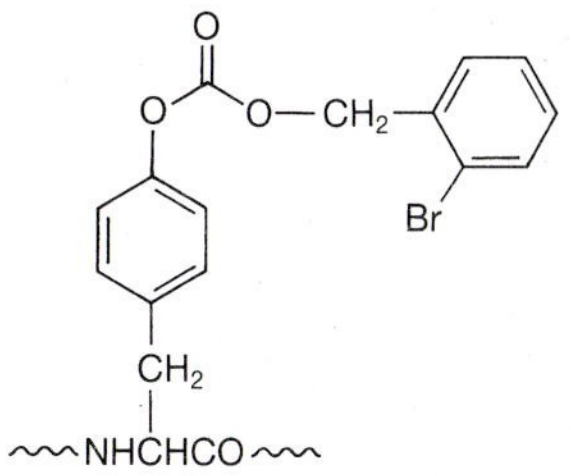

(**4**) Tyr[Z(2Br)]

6.7 Asparagine and glutamine

Amide side-chains have more often than not been left unprotected until recently, despite the fact that unmodified asparagine and glutamine residues often make for poor solubility and side-reactions. Most methods for the strong activation of acylasparagines lead to some dehydration of the side-chain (Scheme 6.18), and the activation of an acylglutamine may

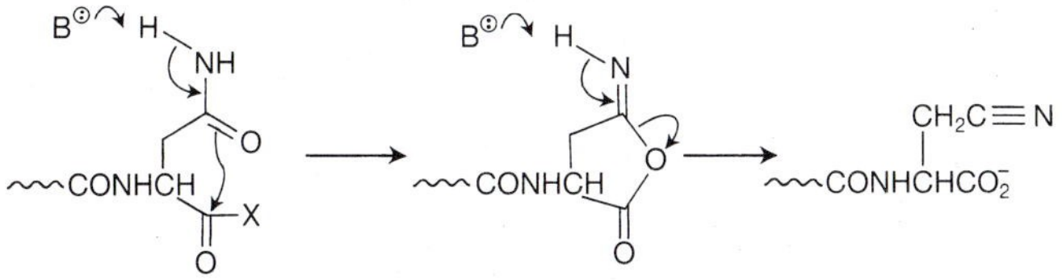

Scheme 6.18.

give a glutarimide (Scheme 6.19). The α-amino group of an *N*-terminal glutamine residue may also react with the side-chain to give a

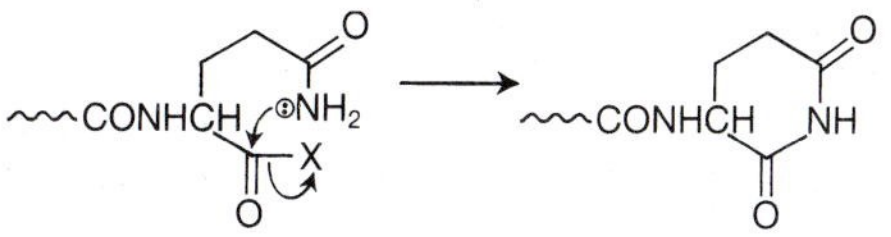

Scheme 6.19.

pyroglutamic acid residue: this is especially liable to occur on Boc removal (Scheme 6.20). Although none of these difficulties is prohibitive,

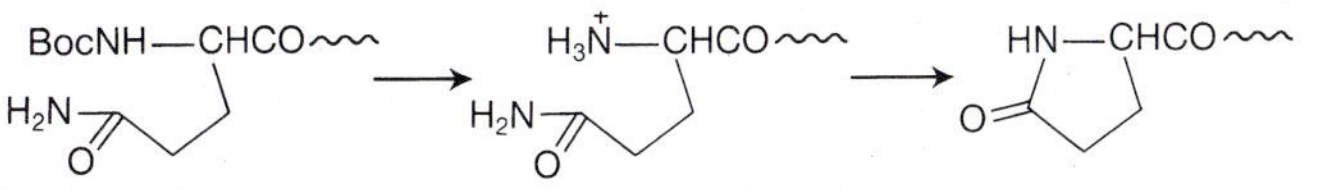

Scheme 6.20. This is a more serious side reaction with HCl/AcOH than with TFA, but may occur to a significant extent even with TFA.

Several important peptide hormones have an *N*-terminal pyroglutamic acid residue (see e.g. Section 8.2.2).

it is clear that there would be advantages in side-chain amide protection (preferably lipophilic and bulky), if this could be simply achieved with a substituent which was amenable to cleavage under mild conditions. The trityl group seems to have properties (stable to base, catalytic hydrogenolysis, and very mild acid, but cleaved by TFA) which are just right for use in conjunction with Fmoc α-protection: preparations of some key asparagine derivatives are outlined in Scheme 6.21 (the glutamine derivatives are prepared similarly).

$$ZAsnOH \xrightarrow{\ i\ } ZAsn(Trt)OH \xrightarrow{\ ii\ } HAsn(Trt)OH$$

$$FmocAsn(Trt)OPfp \xleftarrow{\ iv\ } FmocAsn(Trt)OH \xleftarrow{\ iii\ }$$

Scheme 6.21. Conditions: i, TrtOH/Ac$_2$O/AcOH/H$_2$SO$_4$/50 °C/1.5 h; ii, H$_2$/Pd; iii, FmocOSu/Et$_3$N; iv, DCCI/HOPfp.

6.8   Histidine

The basic and nucleophilic imidazole side-chain does not veto, but creates serious difficulties for, the incorporation of histidine without protection. Racemization occurs in carboxy-activated derivatives through pathways involving heterocyclic nitrogen, to a degree which makes the preservation of chiral integrity a more serious problem during peptide synthesis with histidine than with any other proteinogenic amino acid. Side-chain attack by DCCI has been observed, and acylating agents may lead to the formation of acylimidazoles, which are themselves capable of passing on the acyl group in undesirable ways.

The early literature on the protection of histidine side-chains is confused because the difference between the two heterocyclic nitrogens (**5**) was ignored, arbitrary incorrect assumptions were made about the positions taken up by protecting groups, and the fact that the location of protection might be significant was not appreciated.

The intrinsic reactivity of the two side-chain nitrogens is similar. Unhindered alkylating agents give not only monosubstituted products, but also some disubstitution, as monoalkylation blocks one nitrogen without deactivating the other. The τ- and π-monoalkyl derivatives are typically formed in around 3:1 proportions, but severe steric hindrance in the reagent shifts the balance in favour of the former, so that trityl chloride is regiospecific for the τ-position. Only monosubstitution occurs with acylating and arylsulphonylating agents, because the electron-withdrawing effect of an acyl or arylsulphonyl group sited on one nitrogen is transmitted across the ring and depresses the nucleophilicity of the other. Acetic anhydride is regiospecific for the τ-position, apparently because of thermodynamic control, but hindered or electronically deactivated acylating agents (chloroformates, Boc$_2$O, carbamoyl chlorides) give significant amounts of π-product.

At the time of writing, the best acid-labile protecting group available is τ-trityl **6**. It is very stable to bases and is suitable for use together with Fmoc temporary protection. It is cleaved by aqueous TFA at room temperature, but the π-nitrogen remains free so racemisation is a problem. In conjunction with Boc α-protection, π-benzyloxymethyl (Bom) protection is ideal: it resists TFA, but is cleaved by HF or HBr/AcOH and (less reliably) by catalytic hydrogenolysis: the preparation of BocHis(Bom)OH **7** is outlined in Scheme 6.22.

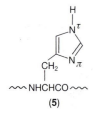

(5)

Suppression of π-basicity and nucleophilicity is necessary to prevent racemisation: this can be done by π-substitution or with a τ-electron withdrawing group.

(6) His(Trt)

Bom cleavage, however, releases formaldehyde; and side-reactions may result unless it is scavenged.

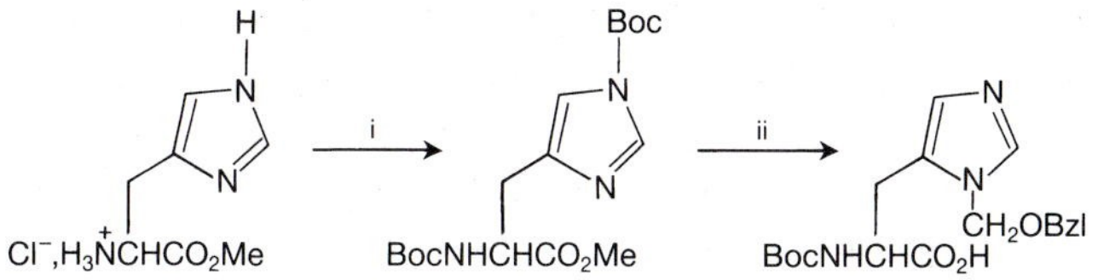

Scheme 6.22.   Conditions: i, Boc$_2$O/Et$_3$N/MeOH (a single crystalline isomer after trituration); ii, BzlOCH$_2$Cl/CH$_2$Cl$_2$, then NaOH/aq. MeOH (a single crystalline isomer).

6.9   Tryptophan

The indole ring of tryptophan does not interfere with peptide bond formation, but captures electrophiles with great ease at several sites, especially the 2-position, and even under relatively mild acidic deprotective conditions competitive scavengers are essential. For protection to be of assistance in diminishing the problem, it is not sufficient merely to block one position—as much π-electron density as possible must be drained from the whole system. The nitrogen is the only ring position for which reversible protection has been investigated. Formyl (For) has been used, but is inconveniently unstable to nucleophiles, and the Boc group is better in that respect: the key intermediate **8** is commercially available. At the final deprotection stage of a synthesis with this intermediate the side-chain deprotection proceeds to the carbamate stage on TFA treatment, but an electron-withdrawing influence remains in place and continues to protect the vulnerable ring while the other protecting groups are being removed. Carbon dioxide loss occurs spontaneously to give the fully deprotected indole side-chain during normal aqueous workup procedures.

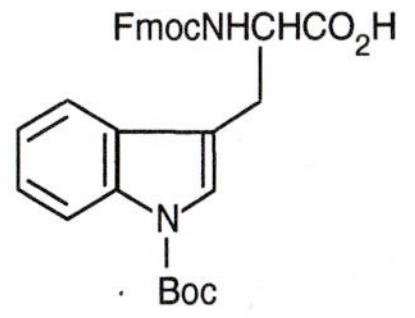

(8)

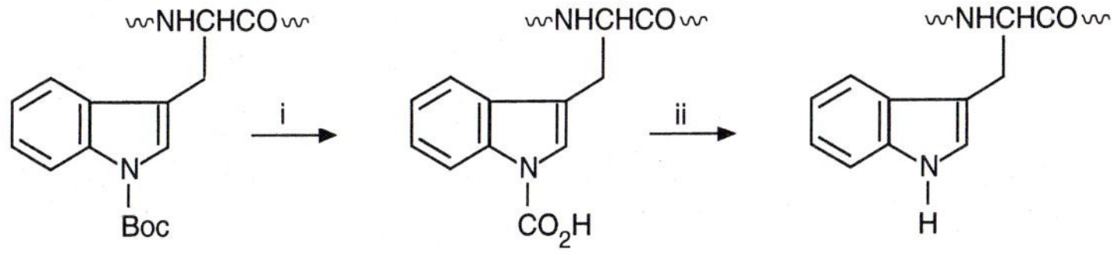

Scheme 6.23.   Conditions: i, TFA; ii, mild aq. acid.

6.10   Proline

The special structure of proline gives it two impediments in synthesis (some steric hindrance to acylation at the secondary amino group; the susceptibility of Xaa-Pro peptide bonds to reduction by sodium in ammonia), but also gives it a resistance to base-catalysed racemization which is unique among the chiral protein building blocks. From a strategic point of view, this is a vital fact, because it allows any coupling from a proline carboxy group, including an acylpeptide fragment coupling, to be performed without racemization.

N-cyclohexyloxycarbonyl (Hoc) indole protection, cleaved by strong acid but stable during the acidolysis of α-Boc, is also valuable, and has been used to great effect by Sakakibara. See the final chapter.

Sodium–ammonia reduction is now rarely used, however.

(9)

An oxazolonium ion
derived from an
activated acylproline

The ordinary oxazolone racemisation mechanism pathway (see Section 5.1.2.2) is not possible with activated acylprolines because they have no NH. Cyclisation to an oxazolonium cation **9** can take place, and such cations are optically labile, but much more vigorous conditions than are employed in peptide synthesis are required. For example, if proline is refluxed with acetic anhydride, an oxazolonium intermediate is generated, and racemic acetylproline is obtained on aqueous workup.

Direct exchange racemisation (see Section 5.1.1.1) via an enolate does not occur in activated acylprolines because unfavorable interactions prevent the formation of the fully planar structure **10** which is needed for simultaneous delocalisation of the enolate and the secondary amide group.

The same conclusion is reached if we consider the other possible conformers (both secondary amide and the enolate could be either *E* or *Z*).

6.11   Glycine, phenylalanine, alanine, leucine, isoleucine, and valine

These amino acids have no side-chain functionality, but peptide synthesis with them can still be problematic. Paradoxically, glycine, achiral and without any side chain at all, is sometimes difficult to deal with, because the unimpeded access to peptide bonds involving it allows side reactions such as hydantoin formation. The same consideration facilitates strong interchain hydrogen bonding, which tends to make peptides with a high glycine content rather insoluble. Of the rest, phenylalanine is usually simple to work with, but may suffer reduction to cyclohexylalanine if extended catalytic hydrogenolysis reaction times cannot be avoided, and the coupling of valine and isoleucine derivatives is relatively slow for steric reasons (which is more than a matter of time and patience: slow coupling enables side-reactions of all kinds to compete more seriously). Only alanine and leucine have a completely clean record.

6.12   Non-proteinogenic residue types

$MeNHCHRCO_2H$

(11)

$H_2NCMeRCO_2H$

(12)

$H_2NCHRCHRCO_2H$

(13)

We exceed our brief slightly, but it is worth mentioning the three general classes **11-13**, which can all be manipulated by the methods used for the proteinogenic amino acids, with only minor differences in experience. Peptides incorporating residues of **11** or **12** are not rare, and several are clinically important. Peptides containing residues of **13** are uncommon, but interesting because it has been shown recently that β-peptide chains can adopt organized conformations in solution, a characteristic formerly thought unique to α-peptides.

7 Strategy and tactics

The last four chapters have presented a summary catalogue of the principal methods by which the various individual reactions and stages of a peptide synthesis may be carried out. It is now necessary to consider how a coherent plan can be drawn up, and the appropriate methods chosen, for a synthesis of many stages.

If we focus on peptide bond formation, and ignore all the other synthetic niceties, we see that a tripeptide can in principle be assembled from amino acids by two strategies, which we will call the $[1 + (2 + 3)]$ strategy and the $[(1 + 2) + 3]$ strategy (Scheme 7.1). These are both

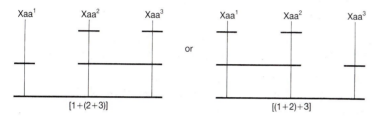

Scheme 7.1.

'stepwise elongation' strategies, in which we start with one residue and add the others one at a time. For a tetrapeptide, however, there are five strategies, of which four are similarly stepwise, but one of which is 'convergent', in that the amino acids are first joined into dipeptide fragments, and these are then condensed together. Scheme 7.2 shows one

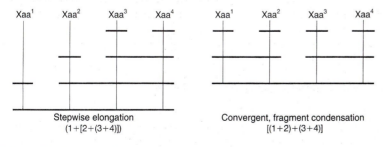

Scheme 7.2. The three other stepwise approaches not shown are $\{[(1 + 2) + 3] + 4\}$, $\{[1 + (2 + 3)] + 4\}$, and $\{1 + [(2 + 3) + 4]\}$.

of the stepwise strategies and the convergent alternative. The verdict of the arithmetic demon on their relative merits is instructive. If there is an 80% yield at each stage, then the $\{1 + [2 + (3 + 4)]\}$ stepwise elongation synthesis will give an overall yield based on residue 4 of 52%, but the convergent fragment condensation strategy will give 64%.

Similar calculations on more extended syntheses argue even more strongly in favour of the 'fragment condensation' approach. The same approach has, in the abstract, two further advantages over the 'stepwise' approach. Firstly, if small fragments are combined to make one which is larger, the differences between the desired product and everything else ought to facilitate its isolation, whereas the addition of residues one at a time would result only in progressive small changes in chemical character and molecular size, with consequently greater isolation and purification problems. Secondly, in a fragment condensation approach, personnel can be assigned each to their fragments to work simultaneously in parallel, but the obstacles of a stepwise synthesis must be negotiated sequentially. There is also an attraction, for certain kinds of programme, in the scope a fragment condensation strategy offers for producing a set of related analogues with variable sequences in one region.

But there are contrary considerations. Fragment condensation is better performed at some positions than others, so it is a constrained strategy, and its protective group tactics are in general more complicated than those of a stepwise synthesis strategy from the carboxy end, which is the preferred stepwise approach.

For a classical 'solution synthesis', the optimum approach is usually stepwise synthesis of fragments for assembly in an overall convergent strategy.

The principal strategic constraint is the desirability of proceeding in such a way as to minimize the risk of racemization throughout. There is no real risk except at the activated amino acid residue during the activation and peptide bond formation stages. Furthermore, if the azide or DCCI/HOBt procedures are used, racemization is usually slight, with small peptides at least; and if the carboxy component is an alkoxycarbonyl, trityl, or Nps amino acid, or an acyl peptide with *C*-terminal proline, racemization is hardly ever significant, whatever the coupling procedure; if the carboxy component is an acylpeptidylglycine, then the activated residue has no chirality to be concerned about. The situation is summarized in Table 7.1. It is clear that, from a racemization point of view, the only acceptable strategies are those in which as many as possible of the peptide bonds in the target are formed as in column A, with the rest being formed as in column B. Even strict adherence to these rules gives no guarantee with fragment condensations, but the traditional view that any racemization would be a disaster without remedy is somewhat belied by the fact that there have been cases where epimeric peptide contaminants arising from azide or DCCI/additive fragment condensations have been separated by countercurrent distribution, a most powerful technique. Advances in separation technology are taking place all the time, and it may be that the traditional view was unduly pessimistic, but it will clearly always be desirable to plan on the basis of minimizing the danger of racemization.

It is not possible to generalize to any great extent about strategic decision-making. There are too many variables. The scale, complexity,

Table 7.1 Racemization risks in practical peptide synthesis* at the peptide bond forming stage**

A Safe	B Usually safe***	C Unsafe
Activation and coupling of any alkoxycarbonyl-amino acid by any conventional method	Activation and coupling of any carboxy component by the azide method****	Activation and coupling of any carboxy component not in column A by any method not in column B
Activation and coupling of acyl-peptidylglycines and acylpeptidyl-prolines by any conventional method	Activation and coupling of any carboxy component by the DCCI/HOBt method or a variant	

Protected amino acids whose α-protection precludes oxazolone formation (e.g. Nps, Trt derivatives) may also be activated and coupled by any method without risk of racemization.

* With proteinogenic amino acids and established methodology.
** There is generally no real risk at other stages, except at the esterification of the first residue to the resin in solid phase synthesis.
*** A few other procedures have some claim to be listed here, but not enough experience has been recorded to admit sound generalization.
**** But not if the coupling is sluggish.

and composition of the target, and also the context in which the target is set, all have to be reckoned with. Di- and tri-peptides can only be made, and tetrapeptides are perhaps best made, by stepwise synthesis. Completely stepwise solution synthesis has been taken beyond twenty residues (e.g. porcine secretin, 27 residues) but from, say, five or six residues upwards, a fragment condensation strategy or (to an increasingly dominant degree) a stepwise solid phase synthesis (see later) is now more likely to be favoured. The number and distribution of glycine and proline residues are important determinants in the detailed design of a fragment condensation strategy, and if there are disulphide bridges the way in which these are to be constructed must usually be considered first and the rest of the synthesis planned around them. If there is a problem-child (Met or Trp in particular) in the target, it will be preferable to introduce it late in the synthesis. It is also prudent to maintain flexibility where possible, so that if it happens that the first chosen order of fragment assembly fails, then a different order of assembly of the same fragments can be tried. Thus the attempted conjunction of four fragments A, B, C, and D by {[A + (B + C)] + D} or {A + [(B + C) + D]} strategies would be unsatisfactory if the BC intermediate proved unreactive or insoluble. In such a case a strategic switch to [(A + B) + (C + D)] or

{[(A + B) + C] + D}, by avoiding the troublesome intermediate, might nevertheless succeed.

And now to protective group tactics. Consider the relatively trivial target GlyLysGly, which is best approached with the stepwise strategy outlined in Scheme 7.3. The minimum requirement for this to be

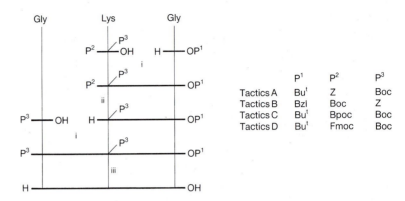

	P^1	P^2	P^3
Tactics A	Bu^t	Z	Boc
Tactics B	Bzl	Boc	Z
Tactics C	Bu^t	Bpoc	Boc
Tactics D	Bu^t	Fmoc	Boc

Scheme 7.3. Conditions: i, coupling; ii, selective deprotection; iii, complete simultaneous deprotection.

workable is that the protecting group P^2 must be orthogonal to the protecting groups P^1 and P^3. P^2 is termed a 'temporary' protecting group, because it is only in position for a brief phase of the synthesis; P^1 and P^3 are called 'permanent' protecting groups because they accompany the residue they protect from its incorporation until the final deprotection. There are many ways of satisfying the orthogonality requirement. Scheme 7.3 illustrates the four most important tactical approaches using well-tried protecting groups (Table 7.2), which can by extension of the same principles be used for more complicated examples. But not all tactics are applicable all the time. Tactics A and B(b), for example, would generally be precluded by the presence of methionine or cysteine in the target.

The discussion has so far been framed in solution synthesis terms, but is equally relevant to the solid phase technique (see later): Scheme 7.3 represents a solid phase synthesis if P^1 is the resin carrier, which can be regarded as the *C*-terminal permanent protecting group, and tactics B(a) and D are in fact the tactics of the Merrifield and Sheppard methods of solid phase synthesis respectively.

The tactical issues are a little more complex in fragment condensation strategies. In general such strategies require, *inter alia*, the stepwise preparation of a number of peptide acids or activatable derivatives with differential protection which enables further elaboration from the amino terminal after fragment condensation. The necessary intermediates may be reached by salt couplings, or through fully protected peptides—which must then have two orthogonal types of temporary protection, as well as permanent protection for the side-chains which is orthogonal to both types of temporary protection. Suppose, by way of illustration, that a

Table 7.2 The principal protective group tactics for stepwise peptide synthesis from the carboxy end

Tactics	A	B	C	D
Temporary α-amino protection	Z	Boc	Bpoc	Fmoc
Conditions for cleaving temporary protection	H₂/ catalyst	Mildly acidic	Very mildly acidic	Basic
Permanent protection	Boc, OBuᵗ, etc.	Z, OBzl, etc.	Boc, OBuᵗ, etc.	Boc, OBuᵗ, etc.
Conditions for cleaving permanent protection	Mildly acidic	(a) Strongly acidic or (b) H₂/ catalyst	Mildly acidic	Mildly acidic

chosen strategy calls for a Lys(Boc)Lys(Boc) dipeptide fragment to be inserted between two other fragments, to make a structure —COLys(Boc)Lys(Boc)NH—. The salt coupling stratagem might be employed here, as in Scheme 7.4, or methyl ester protection could be used

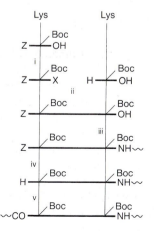

Scheme 7.4. Conditions: i, activation (e.g. DCCI/HOSu to give the succinimido ester); ii, coupling in the presence of base (e.g. DMF/Et₃N); iii, coupling by a racemization-free method (e.g. DCCI/HOBt); iv, H₂/Pd; v, coupling, by a racemization-free method, unless the residue being activated is Gly or Pro.

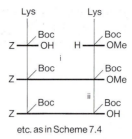

Scheme 7.5. Conditions: i, coupling; ii, saponification.

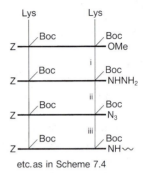

etc. as in Scheme 7.4

Scheme 7.6. Conditions: i, N_2H_4/MeOH; ii, alkyl nitrite ester/H^+; iii, coupling.

as in Schemes 7.5 or 7.6. The reader should by now be able to see several other variations too.

A tactical decision which has so far not been faced in this chapter is whether or not to protect side-chains at all in those cases where it can be regarded as optional (Met, Ser, Thr, Tyr, Arg, Asn, Gln, Trp; in some situations also Asp and Glu). Minimal protection has the attractions of using simpler (hence cheaper) amino acid derivatives as starting materials, and offering fewer opportunities for side-reactions at the final deprotection stage. On the other hand, solubility in organic solvents may become a problem with large numbers of unblocked polar groups, and in repetitive syntheses where there is no purification of intermediates—solid phase methods in particular—the consequences of minor side-reactions can accumulate and become serious. With Met there is uncertainty about whether or not it is advantageous to protect the side-chain, but with all other side-chains maximal protection is current practice.

Other things being equal, the best tactical approach for any synthesis will be that for which the final complete deprotection is performed in a single operation, under the mildest possible and most reliable conditions. Acidolysis (unlike catalytic hydrogenolysis, which can be confounded unpredictably by catalyst poisoning, precipitation of partially deprotected material, and inordinate slowness) rarely fails to proceed according to the rulebook. Side-reactions attributable to electrophilic species generated by the cleavage are a problem, but there have been very through studies which give a partly rational and partly empirical basis for minimizing such difficulties by the addition of scavengers. Tactics A, C, and D are therefore generally preferable to tactics B.

Exercises

Before proceeding to the next chapter, readers might well with advantage consider how the principles outlined in this chapter could be applied to the synthesis of the following.

ProLeuGlyNH$_2$
GlyProTrpMetOMe
BocTrpMetAspPheNH$_2$

One solution for each case can be found in the section on porcine gastrin I (Section 8.2.2), but there are many possibilities.

No attempt is being made in this Primer to cover the synthesis of cyclopeptide structures, in which terminal or side-chain amino and carboxy groups form peptide or side-chain amide links respectively. Such structures occur in diverse important natural products, and are also important in drug design. The principles and reagents involved in their synthesis are the same as for linear peptides, with two special twists. Firstly, it is necessary to manouevre so that only one amino group and only one carboxy group can be exposed at the same time in a late linear intermediate, which imposes constraints on protecting group strategy. And secondly, it is usually necessary to perform the cyclization by coupling the amino and carboxy groups of the late linear intermediate under high dilution conditions, to favour intramolecular cyclization over intermolecular oligomerization. This is a general principle in the formation of macrocycles, unless there is assistance to cyclization by interaction with a template, or fortuitous conformational effects operate. With these considerations in mind, devise a reasonable synthesis of the following, in which the pentapeptide AlaLysGlyGluAla is cyclized with a peptide bond between its terminal groups and an amide bridge between its functional side-chains.

AlaLysGlyGluAla

8 Solution peptide synthesis

8.1 Introduction

Hundreds of peptides and proteins have been synthesized by the so-called 'classical' solution approach to peptide synthesis. The flexibility of the approach makes it impossible to convey its full scope with a few examples, and we must content ourselves with three historically important and spectacular successes.

8.2 Case studies

8.2.1 Oxytocin

HCysTyrIleGlnAsnCysProLeuGlyNH₂

Oxytocin (**1**)

Oxytocin (**1**) is involved in the control of uterine contraction during labour, and of milk release afterwards. It was the first peptide hormone to be synthesized, by du Vigneaud and his collaborators, who reported in 1953, but the methods employed are largely of historical interest only. A later, improved, synthesis by du Vigneaud with Bodanszky is much nearer to present-day practice in both strategy and tactics (Scheme 8.1).

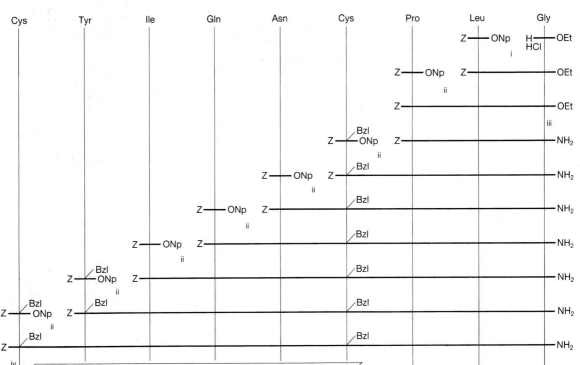

Scheme 8.1. Conditions: i, Et₃N/CHCl₃; ii, HBr/AcOH on the protected amino component, then conversion to the free base form, either before or during reaction with active ester in CHCl₃, EtOAc or DMF; iii, NH₃/MeOH; iv, Na/NH₃(l), then O₂, followed by countercurrent distribution purification.

Oxytocin and its close relative vasopressin have both been subjected to exhaustive structure-activity studies, during the course of which hundreds of synthetic analogues have been synthesized. As an example of an entirely different approach, consider the synthesis of the biologically active [7-(azetidine-2-carboxylic acid)]-analogue (Scheme 8.2). At the

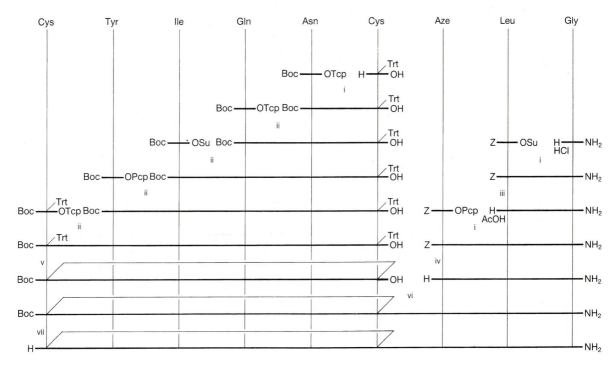

Scheme 8.2. Conditions: i, Et$_3$N/DMF; ii, TFA (with added TrtOH to suppress *S*-detritylation), then active ester/Et$_3$N/DMF; iii, H$_2$/Pd(C)/AcOH; iv, H$_2$/Pd(C)/aq. MeOH; v, I$_2$/AcOH; vi, DCCI/ HOBt/DMF; vii, 90% aq. TFA, then purification by gel permeation chromatography.

time this synthesis was undertaken, the optically pure replacement amino acid was only available in very limited amounts, so it was necessary to introduce it late in the synthesis, using a strategy related to those pioneered by Rudinger and others for peptides of this class.

Azetidine-2-carboxylic acid is the four-membered analogue of proline.

8.2.2   Porcine gastrin I

The gastrins are peptide hormones which can be isolated from the antral region of the stomach. They stimulate gastric secretion. In most mammals there are at least three forms, which are derived from a common precursor, in circulation; these forms may or may not be sulphated at tyrosine, and there are species variations, so there are many gastrins. We shall take the first Liverpool synthesis of porcine gastrin I (**2**)

```
 1          5            10            15    17
GlpGlyProTrpMetGluGluGluGluGluAlaTyrGlyTrpMetAspPheNH₂
```

Porcine gastrin I (**2**)

Refer back to table 7.1 for the 'racemization rules'.

as our example. A fragment condensation strategy was adopted, with three main fragments of comparable size, which were chosen to exclude methionine from the central fragment, and thus permit the employment of hydrogenolysis in its preparation. Fragment conjunction was performed at such points or by such methods as the racemization rules dictated. The strategy used enabled biological testing to be carried out on partial sequences, and led rapidly to the important conclusion that the essential features for activity were all contained in the *C*-terminal tetrapeptide. The synthesis, which is shown in Schemes 8.3–8.6, was not without snags, but it set the style for many others, both at Liverpool and elsewhere.

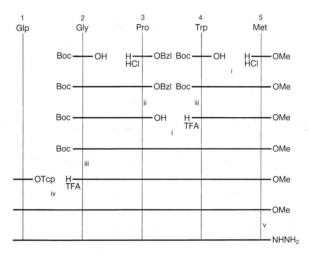

Scheme 8.3. Conditions: i, DCCI, after conversion (Et$_3$N) of the amino component to the free base form; ii, H$_2$/Pd(C)/EtOAc/MeOH; iii, aq. TFA; iv, Et$_3$N/DMF; v, N$_2$H$_4$/MeOH.

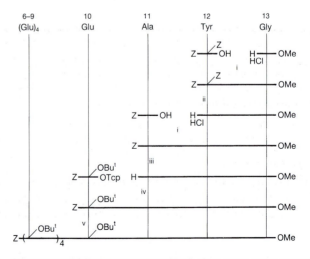

Scheme 8.4. Conditions: i, DCCI, after conversion (Et$_3$N) of the amino component to the free base form; ii, H$_2$/Pd(C)/HCl/MeOH; iii, H$_2$/Pd(C)/MeOH; iv, DMF; v, iii and ZGlu(OBu^t)OTcp/DMF four times.

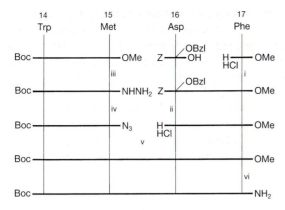

Scheme 8.5. Conditions: i, DCCI, after conversion (Et$_3$N) of the amino component to the free base form; ii, H$_2$/Pd(C)/MeOH/HCl; iii, N$_2$H$_4$/MeOH; iv, HNO$_2$; v, Et$_3$N/DMF; vi, NH$_3$/MeOH.

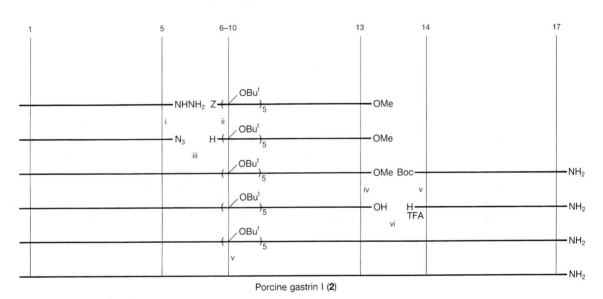

Porcine gastrin I (**2**)

Scheme 8.6. Conditions: i, Bu^{t}ONO/HCl/DMF/−30 °C; ii, H$_2$/Pd(C)/MeOH; iii, DMF/0 °C; iv, NaOH/aq. dioxan; v, TFA; vi, mixed anhydride formation by reaction of the carboxy component with (PhO)$_2$POCl, then addition of the amino component and Et$_3$N. Final purification was by gel permeation chromatography and ion exchange chromatography.

8.2.3 Human insulin

Insulin is involved in the regulation of blood glucose levels: it is used clinically in the control of *diabetes mellitus*. There are minor species variations in sequence. Structures **3** and **4** show the sequences of the

1   10   15   20 21

HGlyIleValGluGlnCysCysThrSerIleCysSerLeuTyrGlnLeuGluAsnTyrCysAsnOH

Human insulin A chain, reduced form (**3**)

```
  1          5          10         15         20         25         30
```

HPheValAsnGlnHisLeuCysGlySerHisLeuValGluAlaLeuTyrLeuValCysGlyGlyGluArgGlyPhePhe- TyrThrProLysThrOH

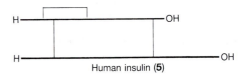

Human insulin B chain, reduced form (**4**)

reduced human A and B chains respectively: **5** shows the whole hormone in diagrammatic form. Bovine insulin was actually the first to be

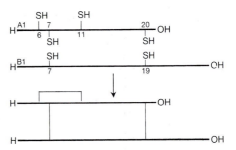

synthesized, by the co-oxidation of separately prepared A and B chains (Scheme 8.7). Unfortunately the dimerization shown in Scheme 8.7 is

Scheme 8.7.   Conditions: mildly oxidative.

only one of 12 possible ways in which the A and B chains can combine, and there are also 4 possibilities for the formation of internally bridged single chains; for A–A and B–B combinations (48 and 2 respectively); and innumerable products comprising more than two chains. Chinese scientists nevertheless managed to define conditions which allowed them to isolate essentially pure crystalline insulin. Their synthesis was not, however, a controlled synthesis, which was only achieved (for the human hormone this time) after methods for the stepwise formation of disulphide bonds had been developed.

Nature does not have this problem, because *in vivo* insulin is derived from proinsulin, a single chain precursor in which the A and B sequences are linked by a connecting peptide. Proinsulin spontaneously adopts a conformation in which the disulphide links form correctly, after which enzymic excision of the connecting peptide gives insulin.

   The first completely controlled total synthesis of human insulin is sketched in broad outline in Schemes 8.8–8.10: its key features are the selective removal of a trityl group in the presence of Bpoc and other acid-labile groups (a nearly maximal protection approach was adopted, with all hydroxy, amino, and carboxy groups blocked by *t*-butanol-derived protection; histidine, arginine, and amide side-chains were unprotected);

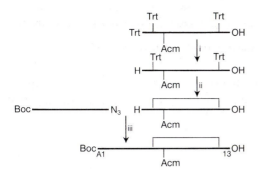

Scheme 8.8. Conditions: i, AcOH; ii, I$_2$/TFE; iii, DMF, followed by countercurrent distribution purification.

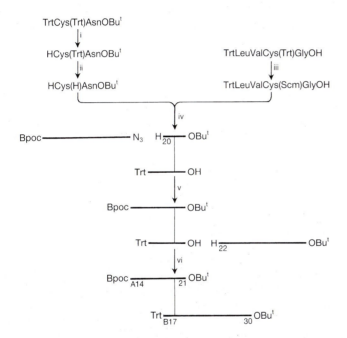

Scheme 8.9. Conditions: i, AcOH; ii, Hg(OAc)$_2$; iii, ScmCl/Et$_2$NH; iv, CHCl$_3$-MeOH; v, DMF; vi, DCCI/HOBt/DMF, followed by countercurrent distribution purification.

the use of the sulphenyl thiocarbonate method to make the unsymmetrical two-chain disulphide starting material; and the selective formation of a disulphide bridge between two Cys(Trt) residues in the presence of Cys(Acm), leaving the latter free for the separate, final disulphide ring closure.

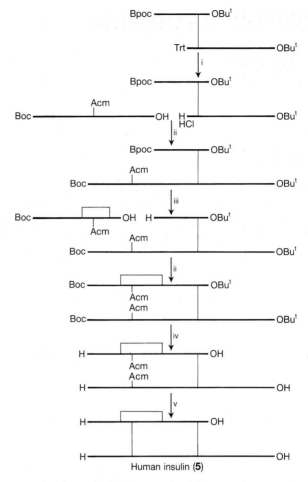

Human insulin (**5**)

·**Scheme 8.10.**   Conditions: i, HCl/TFE, apparent pH 3.5, brief; ii, DCCI/HOBt/DMF; iii, 90% aq. TFE/60 °C/2 h; iv, 95% aq. TFA; v, I$_2$/AcOH, followed by countercurrent distribution purification.

The human insulin used clinically is not manufactured by this route, however: most of it (several tons a year) is made by recombinant DNA technology, and the balance is produced by enzymatic modification of the porcine hormone.

The cases of insulin and ribonuclease are not limiting cases, either in terms of size or complexity, but the many protein syntheses achieved in more recent years have mostly employed strategies in which solid phase techniques (see next chapter) are used to reach oligopeptide intermediates which are then assembled in homogenous solution (as for GFP, see final chapter), either by analogy with the methods exemplified in this chapter, or by ligation procedures.

9 Solid phase peptide synthesis

9.1 Introduction

The beautifully simple idea of solid phase peptide synthesis was first detailed by Merrified in 1963, in a publication which, unusually for an important modern scientific paper, bore his name as sole author. The principle is outlined in Scheme 9.1. The term 'solid phase' is actually

J. Amer. Chem. Soc., 1963, **85**, 2149

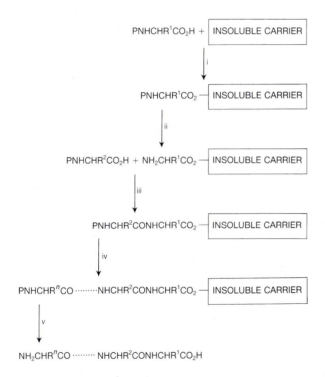

Merrifield was awarded a Nobel Prize for this seminal invention, in 1984.

Scheme 9.1. Conditions: i, attachment of a protected amino acid to the polymeric carrier; ii, N(α)-deprotection; iii, coupling of another protected amino acid to the aminoacylopolymer; iv, repetition of ii and iii as appropriate; v, cleavage of the peptide–polymer link and deprotection of the peptide.

misleading: the reactions do not take place in, or on the surfaces of, solid phases which are in heterogenous contact with solutions, but in swollen gel systems produced by the penetration of solvent and solute molecules right into the polymeric matrix. Every facet of every stage of Scheme 9.1 has been subjected to detailed scrutiny with a view to optimization.

Instrumentation has been developed to enable the repetitive parts to be carried out fully automatically, with delivery of measured quantities of the appropriate reagents, protected amino acids, and solvents according to predetermined protocols. Once set up and started, such automated 'synthesizers' need neither the intervention nor the presence of any operator. Less sophisticated systems comprising the plumbing but not the programming facilities are available for convenient 'manual' solid phase synthesis by an operator. These have the advantage that the operator can keep an eye on things and take remedial action if failures of chemistry or technology are detected, but the rapid developments in automated monitoring which are taking place will probably eventually enable feed-back control without an operator even for non-routine cases. The whole area has become a booming multimillion dollar business, with numerous international companies competing to supply specialized equipment and consumables. Small-scale academic applications get most space in the literature, but there are important large-scale industrial applications of solid phase synthesis too, although most industrial peptide production still depends on classical solution methods. In the necessarily superficial account of solid phase peptide synthesis which follows, we shall confine ourselves to the chemistry involved, but we should remember all the time that there is a sophisticated supporting technology.

As a means to an end, the approach was welcomed with enthusiasm by biochemists and others who were impatient for peptides, but who were perhaps not much interested in the synthetic chemistry *per se*. In part as a reaction to this, those of a more chemical inclination were at first very critical and patronizing. There was a feeling about that organic synthesis without the meticulous characterization of intermediates was bad science, and not really cricket. But solid phase methods now dominate synthetic peptide research, and the principle has been successfully extended to other fields as well, especially peptide sequence analysis and oligonucleotide synthesis. For peptide synthesis, Merrified and others have developed his original chemistry to a fine art (Section 9.2 opposite), while Sheppard and his school have taken Merrifield's

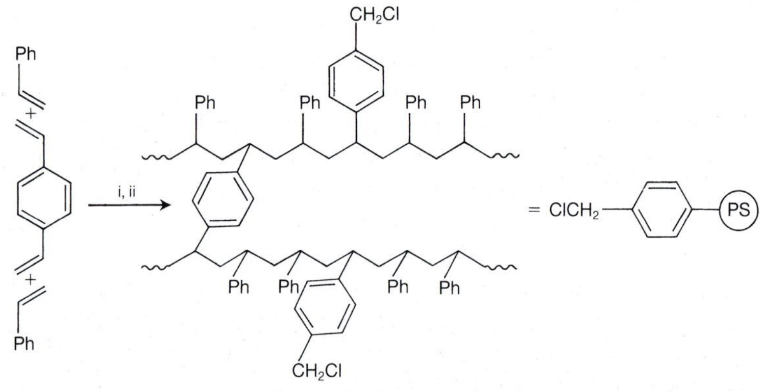

Change is afoot here, however.

Scheme 9.2. Conditions: i, copolymerization (*c.* 2% divinylbenzene), ii, MeOCH$_2$Cl/SnCl$_4$.

basic concept and remodelled it very successfully with alternative chemistry (Section 9.3 on p. 75).

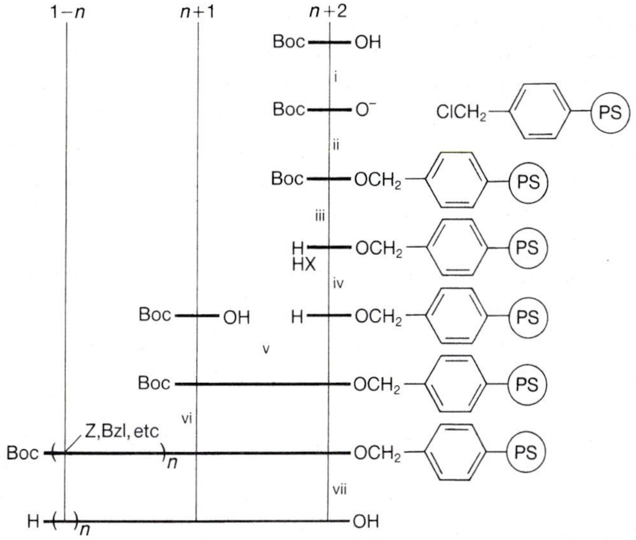

Scheme 9.3. Conditions: i, salt formation; ii, esterification by nucleophilic displacement; iii, deprotection with mildly acidic conditions; iv, neutralization; v, DCCI; vi, repetition of iii–v as appropriate; vii, vigorous acidolysis.

9.2 The Merrifield approach

9.2.1 Principles

A year after his preliminary publication establishing the solid phase principle, Merrifield described a synthesis of bradykinin (see Section 9.2.2.1) which has served as the prototype for the standard approach associated with his name. In this approach, polystyrene crosslinked by the incorporation of a small amount of divinylbenzene is chloromethylated (Scheme 9.2), and the target peptide *C*-terminal amino-acid residue, *N*-α-protected with a Boc group, is attached to the polymer by nucleophilic displacement. Deprotection with mild acid followed by neutralization gives a free amino group, to which the second amino acid residue can be coupled by means of DCCI. The deprotection, neutralization, and coupling stages can be repeated with appropriate Boc amino acids until the target sequence is fully assembled. Side-chains are blocked with permanent protecting groups derived from benzyl alcohol, or other suitable groups which resist the mild acid required for removal of the temporary α-Boc protection. At every stage, after the attachment, deprotection, neutralization, or coupling reaction, the completely insoluble polymer–peptide conjugate is washed exhaustively to remove excess reagents and coproducts. Finally, treatment with strong acid simultaneously strips off all the protecting groups and cleaves the benzyl ester polymer–peptide link (Scheme 9.3), giving the crude required peptide. From the attachment of the first residue to the addition of the last, the peptide–resin conjugate remains in the same vessel, retained there by the sintered glass partition through which all the washings pass to waste.

The main problems encountered relate to (a) non-quantitative reactions; (b) incomplete orthogonality between the temporary blocking groups and the permanent blocking groups (which include the peptide-polymer link, cleavage of which results in solubilization and loss of peptide); and (c) side-reactions, especially at the final complete deprotection and cleavage from the carrier.

If any *N*-terminal residue of the peptide–polymer conjugate remains unacylated after any coupling stage, this will lead to an end product which is contaminated with incomplete sequences. The unreacted residue may be successfully coupled in a subsequent cycle, in which case the resulting contaminant is a deletion peptide; if it is not, then a truncated target sequence results. Elementary arithmetic leads to the conclusion that near-quantitative reactions are essential. For a synthesis of 10 cycles, even if each coupling stage is 99% complete and there is no termination of chain growth to give truncated sequences, the end product is only about 90% pure target peptide, the 10% being largely made up of the 10 possible single-residue deletion peptides; for 100 cycles, the end product will be about 36% pure, with 38% of single-residue deletion contaminants (of which there are 100 possibilities), 18% of two-residue deletion contaminants (of which there are some 5000 possibilities), the bulk of the remainder being made up of three- or four-residue deletions (of which there are over 4 000 000 possibilities).

Early critics maintained that these calculations exposed a hopeless situation: the level of completeness required to give purifiable end peptides was beyond attainability. This view can now be seen to have been too dismissive, on three counts.

ninhydrin

Firstly, in favourable cases, coupling can in fact be pushed to very nearly 100% by use of excess acylating agent, or performing the coupling stages more than once; the use of highly activated intermediates like symmetrical anhydrides may be advantageous. Completeness can be monitored by colour tests for residual free amino groups—the Kaiser test, based on the reaction of peptide–polymer samples with ninhydrin, has been the standard method. Unfavourable cases, where a residue on the polymer proves resistant to coupling, are often due to aggregation or ordering of the peptide chain concealing the group to which access is necessary; now that there is some understanding of this kind of problem, it can be eased by taking deliberate steps (such as adding salts) to disrupt undesirable interactions. If despite such measures the amino groups of a few chains fail to couple, the consequences can be limited by 'capping' with acetic anhydride to terminate the further growth of these chains and prevent the appearance of single-residue deletion peptides in the end product. Some protocols involve performing double couplings and a capping operation in every residue cycle to be on the safe side.

Secondly, it was a premise of the early critics that separation techniques would never be able to cope with the kind of purification problem that the solid phase approach would pose with peptides of any size. In the meantime, however, advances in separation and characterization technology have been far greater than could have been anticipated.

HPLC with on-line electrospray mass spectrometry to identify peptides as they emerge, for example, is an enormously powerful technique applicable to tiny amounts.

Thirdly, experience has shown there are in any case many applications for synthetic peptides where inhomogenous preparations, provided they are used critically, can nevertheless give valuable results. A striking example is Geysen's approach to the investigation of antigenic determinants ('epitope mapping'), in which large numbers of related peptide sequences are assembled simultaneously by parallel multiple Merrifield synthesis on batteries of polyethylene rod supports, each of which has its own reactor well. The peptides are assembled on aminoalkyl appendages which are anchored to the rods by graft polymerization; they are deprotected and investigated for antibody binding whilst still attached to the rods. All the probes thus made must be inhomogenous, but the differences between their principal components still show up, enabling rapid screening.

In view of the initial criticisms levelled at the solid phase peptide synthesis approach, it is ironic that deliberately generated mixtures have also proved valuable. If mixed amino acid derivatives are used in carefully thought-out combinations, 'peptide libraries' result which enable oligopeptide sequences to be screened in vast groups. This makes it possible to home in on biologically active structures without the need to carry out large numbers of individual syntheses, as in a classical structure–activity study. 'Combinatorial synthesis', like the solid phase concept itself, has also been exported from peptide synthesis to other sectors of bioorganic chemistry.

See *Combinatorial Chemistry*, in the Oxford Chemistry Masters series.

In the standard Merrifield approach, the differentiation between the temporary α-amino protection on the one hand, and the permanent side-chain protection and peptide–resin link on the other, is dependent on the much greater acid-lability of Boc groups than benzyl ester and related groups. The difference is perfectly sufficient for syntheses involving a modest number of steps, but slight side-chain deprotection of Lys(Z) residues can occur under Boc-cleavage conditions, leading to the formation of branched peptides; and some loss of peptide from the peptide–resin conjugate takes place at each Boc-deprotection, because the benzyl ester anchor is not entirely resistant to acidolysis. These problems can be met by tuning the acid-lability of the bonds concerned with appropriate substituents. For example, the loss of side-chain protection from lysine under Boc-acidolysis conditions can be greatly reduced by the use of Lys[Z(2Cl)] instead of Lys(Z), and the loss of peptide from the conjugate can be diminished by attaching the first residue through a benzyl ester link of electronically moderated acid-lability, as with 4-hydroxymethylphenylacetamidomethylpolystyrene carriers, or PAM resins (Scheme 9.4). There is a chemical balancing act to be performed here, because the very modifications which endow side-chain protection and peptide–polymer connections with greater stability necessarily require more severe final acidolysis, and consequently greater risk of side-reactions at that stage, which is problematic for this reason in any case. The original HBr/TFA reagent has been largely superseded by liquid HF, despite the serious hazards and technical difficulties associated with it (commercial services for the performance of HF cleavage are on

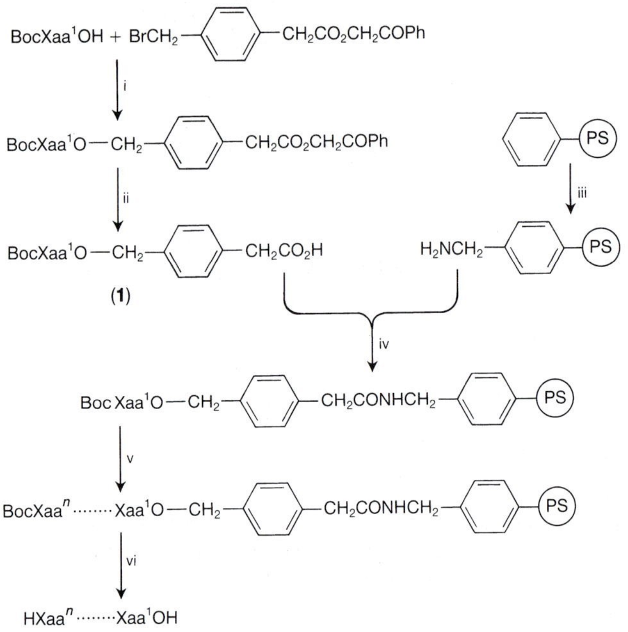

Scheme 9.4. Conditions: i, base/DMF; ii, Zn/aq. AcOH; iii, *N*-hydroxy-methylphthalimide/H$^+$, then N$_2$H$_4$/reflux; iv, DCCI; v, (*n* − 1) cycles of mild acid *N*(α)-deprotection, neutralization, and coupling; vi, deprotection and cleavage from the carrier by vigorous acidolysis.

CF$_3$SO$_3$H
TFMSA

offer), and, as yet less well tried, trifluoromethanesulphonic acid (TFMSA). Scavengers are essential, and a great deal of effort has been invested in the definition of optimum conditions. In particular, it is found best to execute HF cleavage in two stages. In the first stage, HF diluted with substantial amounts of dimethyl sulphide is used to cleave the more labile benzyl–oxygen bonds, under what are presumed to be S$_N$2-promoting conditions. In the second stage, neat HF is used to cleave groups which survive the first stage—Arg(Tos), Arg(NO$_2$), for example. The point of this so called 'low-high' HF procedure is that it avoids the exposure of the peptide to the benzyl cations which would result from a one-step neat HF treatment, since neat HF favours S$_N$1-fission of benzyl–oxygen bonds.

So far, we have assumed that the objective is a peptide with both termini free of substitution, but there is an important variation to note. Many biologically active peptides have an amide function at the *C*-terminal. For the synthesis of these in the oxytocin and vasopressin field, ammonolysis has been used with considerable success to cleave benzyl ester peptide–resin links. Carriers for more general use have also been developed: in these, the first residue is attached through an amide link which is designed to undergo polymer-nitrogen fission on strong acidolysis at the end of the synthesis (Scheme 9.5).

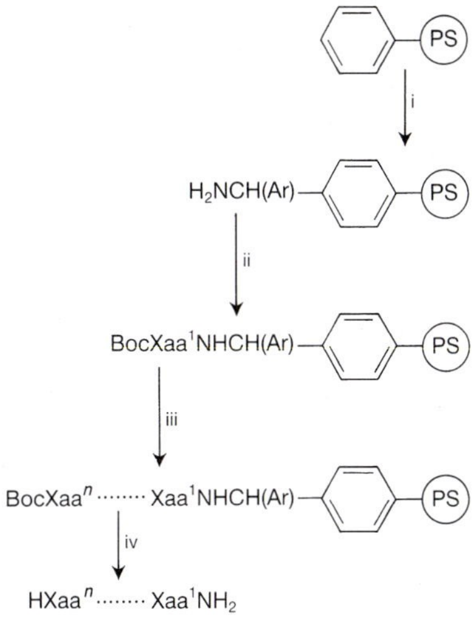

Scheme 9.5. Conditions: i, $ArCOCl/AlCl_3$, then $HCO_2NH_4/HCONH_2$; ii, coupling; iii, $(n - 1)$ cycles of mild acid deprotection, neutralization, and coupling; iv, deprotection and detachment from the carrier by cleavage of the NH-CH(Ar) bond by vigorous acidolysis.

9.2.2 Case studies

9.2.2.1 Bradykinin

Bradykinin (**1**) is a nonapeptide which has an important role in the regulation of kidney vascular blood flow. Merrifield's solid phase synthesis of it (Scheme 9.6) was a turning point in the development of peptide synthesis methodology. The overall yield of chromatographically pure and fully active bradykinin, reckoned from $BocArg(NO_2)$-polymer, was 68%, and the time required was eight days (it would be even less now). Solution synthesis would have taken weeks then, as now.

ArgProProGlyPheSerProPheArg

Bradykinin (**1**)

9.2.2.2 Bovine pancreatic ribonuclease A

The solid phase synthesis of bovine pancreatic ribonuclease A, which has 124 residues and 4 disulphide bridges, by Gutte and Merrifield, was first announced in 1969: refinements and details of the work appeared in 1971. Essentially the same methods as had been used for bradykinin were employed, with HF for final cleavage from the resin and side-chain deprotection. It was not possible to isolate completely pure synthetic enzyme with the separation methods then available, but 0.4 mg material with a specific activity of 78% was eventually obtained. There was no protective differentiation between the SH groups, and oxidation to give disulphide bridges was therefore uncontrolled, but this is a favourable case for it to take place spontaneously to give the natural structure. This work demonstrated convincingly, to all but a shrinking residue of determined sceptics, the power and potential of the solid phase approach. It had taken only a decade for the concept to be brought to this fruition, which also provided the first proof of an important fundamental principle, namely that the natural three-dimensional structure of a protein is determined by the amino acid sequence alone. This principle

A classical solution synthesis of this enzyme was reported by Yajima and Fujii ten years later.

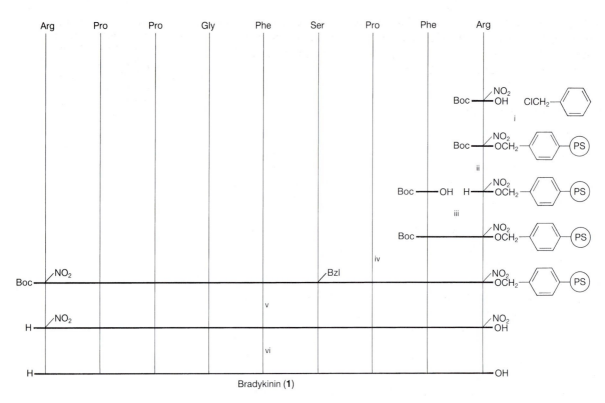

Scheme 9.6. Conditions: i, Et$_3$N/EtOH/80 °C/48 h; ii,, HCl/AcOH, then Et$_3$N/DMF; iii, DCCl/DMF; iv, seven similar cycles of deprotection, neutralization, and coupling with the appropriate Boc amino acids; v, HBr/TFA; vi, H$_2$/Pd(C)/MeOH, followed by ion exchange chromatography purification.

was universally believed (if it was not so, what did control three-dimensional organization, and how could the information be stored and transmitted?), but the experimental evidence up to that time fell short of complete proof. Proteins had been structurally disrupted and deactivated, and then shown to reorganize (in certain favourable cases) into active structures spontaneously on removal of the disorganizing influences. But such experiments were open to the criticism that the disorganized forms might retain some structural information in organized regions, which could survive to nucleate regeneration of the correct active structures. No such objection could be levelled at the formation of enzymically active ribonuclease by total synthesis, since the only structural information put into its construction was the amino acid sequence.

The main chemical problems encountered by Merrifield and Gutte were due to (a) incomplete coupling; (b) loss of side-chain protection; (c) loss of peptide from the resin (more than 1% at each stage); and (d) side-reactions caused by HF. There have been advances on all these aspects in the intervening 20 years, and the synthesis of large peptides by the Merrifield approach, while by no means routine, is now not so daunting. The main differences are the use of faster and cleaner coupling procedures, side-chain protection of greater stability, PAM resins, and moderated HF final deprotection/cleavage. Recent examples include

HIV-1 processing protease (99 residues; the synthetic enzyme was required for the development of inhibitors as potential AIDS drugs; its structure has been determined by X-ray crystallography; numerous close analogues have also been synthesized).

9.3 The Sheppard approach

9.3.1 Principles

In 1971 Sheppard argued that it would be desirable in solid phase synthesis for the polymeric carrier to be chemically similar to the growing peptide chain, so that both carrier and peptide could be well solvated by dipolar aprotic solvents, to give an open gel system in which there was free access of reagents to reactive sites. This was in contrast to the situation with peptides attached to Merrifield cross-linked polystyrenes, where the favoured solvent for swelling the polymeric matrix (CH_2Cl_2) was unable to solvate the conjugated peptide chain well and thereby inclined to encourage its aggregation (depending on the sequence and side-chain protection), reducing the availability of reactive groups. This line of thought led to the introduction of polyamide carriers into solid phase peptide synthesis in place of polystyrenes. Although there has been a great deal of chemical and technological cross-fertilization between the two methodologies, they have developed in parallel, and remain distinct approaches: that associated with Merrified employing polystyrenes, usually with Boc α-protection; and that associated with Sheppard employing polyamides, usually with Fmoc α-protection (hence the common expression 'Fmoc polyamide synthesis').

The standard Sheppard carrier is derived from a cross-linked polyacrylamide incorporating a number of sarcosine methyl ester side-chains ('Pepsyn': Scheme 9.7); the side-chains are extended by treatment

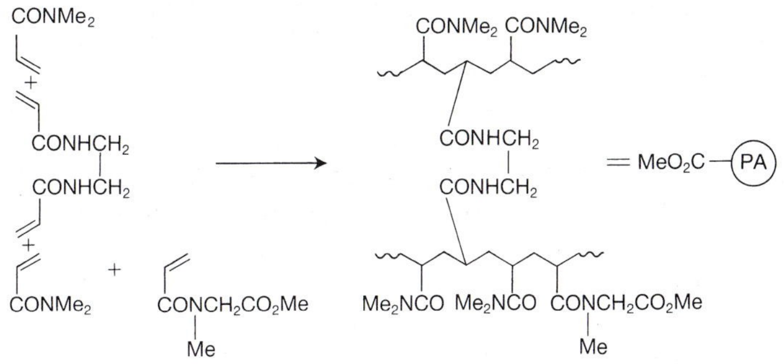

One of the analogues was assembled entirely from protected D-amino acids, giving a protein which folded into an enzymically active complex. This was enantiomeric with the natural enzyme by physical criteria, and and showed reciprocal chiral substrate specificity.

Scheme 9.7. Conditions: carefully controlled suspension copolymerization of dimethyl-acrylamide with about 5% molar proportions of the other monomers, in dichloroethane/DMF/H_2O at 50 °C, in the presence of cellulose acetate butyrate and ammonium persulphate.

with ethylenediamine, followed by coupling with Fmoc-norleucine, which provides an internal reference amino acid (Scheme 9.8). A 'linker' group or resin 'handle' is added after Fmoc cleavage, and the first protected residue of the peptide target is attached to the exposed hydroxy function

The prefix *nor* signifies removal of a side-chain methyl group in amino-acid nomenclature, so norleucine (Nle) has side-chain -$CH_2CH_2CH_3$, cf. leucine -$CH_2CH(CH_3)_2$. The point of employing Nle as an internal reference here is twofold: it does not occur in ordinary peptides, and it is well separated from all the common proteinogenic amino acids on amino-acid analysis.

on the linker. Alternate deprotection and coupling of Fmoc residues with acid-labile side-chain protection assembles the required sequence, which is then α-deprotected and detached from the carrier (Scheme 9.8) with simultaneous side-chain deprotection, by mild acidolysis.

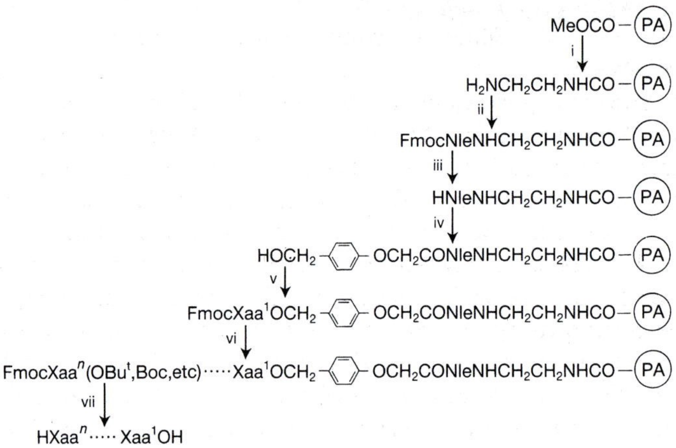

Scheme 9.8. Conditions: i, $NH_2CH_2CH_2NH_2$; ii, $(FmocNle)_2O/DMF$; iii, 20% piperidine/DMF; iv, 4-hydroxymethylphenoxyacetic acid trichlorophenyl ester/DMF; v, $(FmocXaa)_2O/DMAP/DMF$; vi, $(n-1)$ cycles of deprotection with 20% piperidine/DMF and coupling with e.g. $(FmocXaa)_2O/DMF$ or FmocXaaOPfp/HOBt/DMF; vii, 20% piperidine/DMF (if the last residue is introduced with α-Boc protection, this step may be saved), then TFA.

The acid-lability of the peptide–carrier link in Scheme 9.8 is due to the fact that it is a 4-alkoxybenzyl ester; analogously prepared peptide–polymer links cleaved by other means are also available—e.g. **2** cleaved

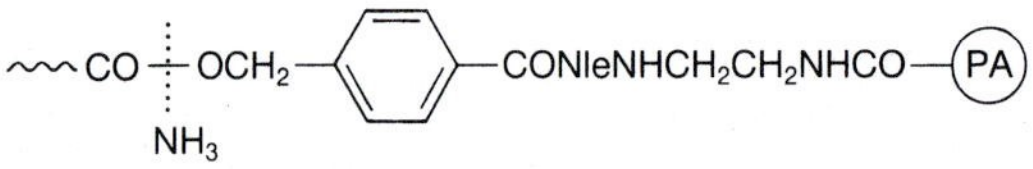

(2)

by ammonolysis and **3** cleaved by acidolysis, to give *C*-terminal amides in both cases. Symmetrical anhydrides or very highly activated esters are

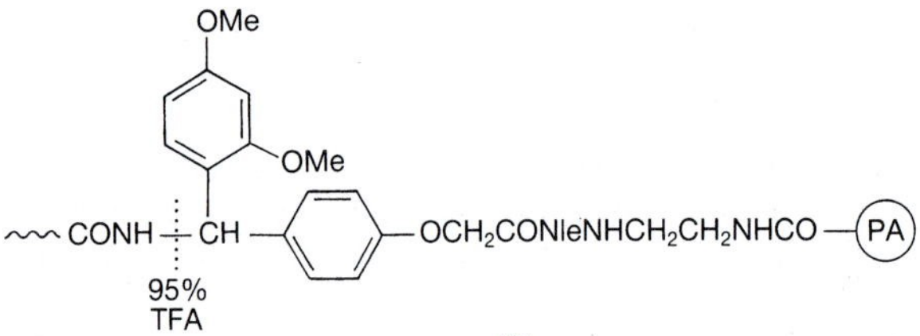

(3)

preferred for the coupling stages: very high levels of incorporation are attainable, even with sequences recognized as difficult in standard Merrifield synthesis. The α-Fmoc, ω-mild-acid-labile protecting group tactics normally used differ with advantage from the α-Boc, ω-strong-acid-labile tactics of the standard Merrifield approach in two ways: (a) the α/ω-protection is absolutely orthogonal as opposed to nearly orthogonal; and (b) the final deprotection can be carried out under much gentler conditions. Further, a single solvent, DMF, is generally used at all stages, reducing the number of wash operations needed, and avoiding difficulties due to swelling and contraction of the polymer with solvent changes. A range of manual and automatic synthesizers appropriate for this approach is available, and rapid technical advances which are outside the scope of this book are taking place, especially in polymeric carriers and supporting systems, in instrumentation for continuous flow operation of the synthesis, and in monitoring with feedback control.

We have alluded several times to the common and serious problem in solid phase peptide synthesis which arises from the tendency of certain sequences to self-aggregate on the resin. This reduces the accessibility of the *N*-terminal group to reactants in solution, which interferes with coupling, deprotection, and monitoring methods based on end-group detection. The problem usually appears, if it is going to, after six or seven residues have been added, when it becomes necessary to use extended reaction times, repeated reaction cycles, hyper-reactive acylating agents, or aggregation-disrupting media, in order to continue chain-assembly. It can be a losing battle. There is now a measure of understanding of the phenomenon, and it is recognized that some kinds of sequence and residue (and hence side-chain protection) are more prone to it that others. But it cannot be predicted with confidence when it will be encountered and when not. Nor would we expect to be able to predict this at the present stage of conformational understanding in peptide and protein chemistry. It is clear enough, however, that backbone to backbone hydrogen-bonding is a major factor, and that it cannot take place if there is *N*-substitution at intervals. Reversible blockade of occasional peptide bond nitrogens is therefore desirable. This is not as easily achieved as it sounds, because *N*-substitution militates against efficient coupling, and adds to the final deprotection. An extremely ingenious solution has been found by Sheppard and Johnson, who recommend the incorporation of an *N*-Hmb-residue at frequent intervals by coupling a bis-Fmoc-Hmb-amino acid. At the next coupling stage after removal of the two Fmoc groups, there is an assisted pathway for acylation which overcomes the steric hindrance (Scheme 9.9).

An alternative solution is to introduce an FmocXaaSerOH dipeptide unit as an acetone acetal type adduct of structure

Fmoc removal and further coupling and deprotection cycles give a sequence containing a pseudoproline residue, which confounds any tendency to conformational organization during assembly, but reverts to a Ser residue on final acidolysis. This works for the Thr case too.

A bis-Fmoc-Hmb-amino-acid

An Hmb-peptide

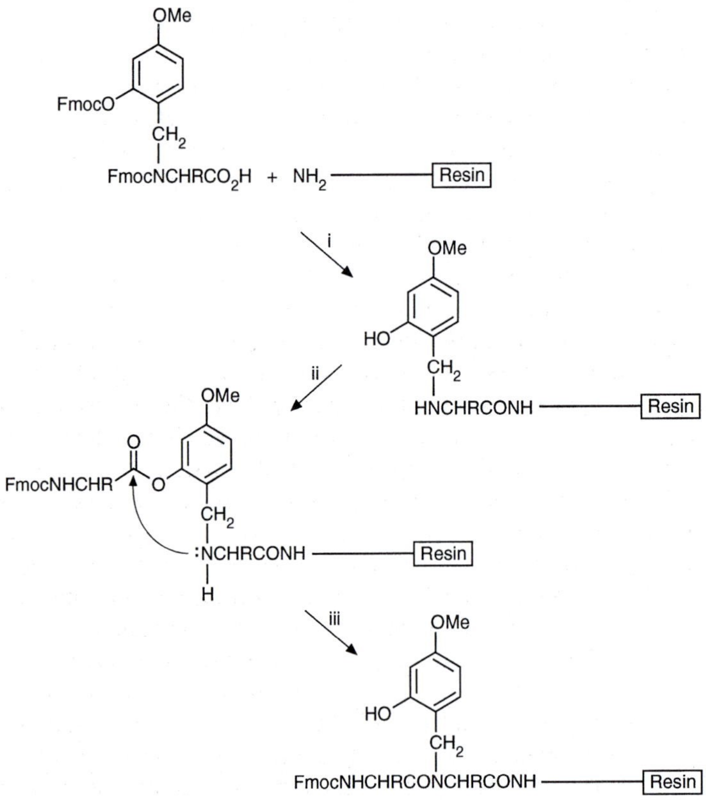

The intramolecular acyl transfer is facile because it proceeds through a six-membered transition state.

Scheme 9.9. Stages: i, coupling and deprotection as usual; ii, acylation at oxygen, the least hindered site; iii, intramolecular acyl transfer from oxygen to nitrogen.

At the end of the assembly the Hmb group can be removed by TFA treatment, along with the other acid-labile groups and release from the resin.

9.3.2 Case Studies

9.3.2.1 Acyl carrier protein-(65–74)-decapeptide

This peptide (**4**) is of interest because its synthesis failed using early Boc-polystyrene techniques, probably because of aggregation-related

65   74
HValGlnAlaAlaIleAspTyrIleAsnGlyOH

Acyl carrier protein-(65–74)-decapeptide (**4**)

phenomena. Sheppard and his colleagues have synthesized it successfully many times, using it as an exercise to test refinements in their methodology. The synthesis summarized in Table 9.1 exemplifies the use of pentafluorophenyl ester activation.

Table 9.1 A solid phase synthesis of acyl carrier protein-(65-74)-decapeptide 4

Carrier	As shown in Scheme 9.8
$N(\alpha)$-Protection	Fmoc
Side-chain protection	Asp(OBu^t), Tyr(Bu^t)
Attachment of the first residue	(FmocGly)$_2$O/DMAP/DMF
$N(\alpha)$-Deprotection	20% piperidine/DMF
Coupling	FmocXaaOPfp/DMF, with catalysis by HOBt in some cases
Cleavage from the carrier and side-chain deprotection	Fmoc removal as usual, then 95% aq. TFA. The crude product had a good amino acid analysis, and was essentially homogenous by HPLC

9.3.2.2   *β-Amyloid (1–43) peptide*

It is a sad but inevitable fact that a high proportion of the lively young people who will, one hopes, read this little book, will lose their reason to Alzheimer's disease in their declining years. Unless, that is, an understanding of the condition enables preventative medicine to intervene. It is at present an inscrutable syndrome, but it is recognized that the presence of extracellular insoluble proteinaceous deposits (known as amyloid plaques) in the brains of stricken individuals is a pathological characteristic. Attention has consequently focused on these plaques, the major constituent of which is a 42- or 43-residue peptide designated βA4. The synthesis of the 43-residue isoform βA4(1–43) is therefore of special interest, as it should open up possibilities for studying it and analogues. It is an excruciatingly difficult synthetic target, because *C*-terminal sequences show a high tendency to aggregate and frustrate chain assembly, and the end product is exceedingly insoluble, making purification difficult. Johnson has recently cracked both problems by use of the Hmb backbone protecting group. During assembly of the sequence, the Hmb group inhibited aggregation. Acetylation of the Hmb groups at their phenolic oxygens while the protected peptide was still attached to the carrier enabled them to survive TFA side-chain deprotection and detachment from the resin: the penta-AcHmb derivative released was soluble to an extent of 10% w/v in 0.1% TFA-MeCN, in which the fully deprotected peptide is virtually insoluble. Purification by chromatography was possible at this stage. Hydrazinolysis then removed the *O*-acetyl substituents and restored the acid-lability of the Hmb groups, which could then be stripped off with TFA. Although this synthesis, which is summarized in Table 9.2 and Scheme 9.10, was one of the first to use backbone blockade and will no doubt be improved on, it was a remarkable achievement with an intractable peptide.

See Johnson, T., *et al., J. Org. Chem,* 1994, **59**, 1745 and *J. Chem. Soc. Perkin Trans I,* 1995, 2019

Table 9.2   Solid phase assembly, β-amyloid (1-43) peptide

Starting material	FmocThr(Bu^t)-Pepsyn KA, i.e. FmocThr(Bu^t) esterified to a 4-alkoxybenzyl alcohol, which was linked through the alkoxy group to a norleucine residue (as an internal standard for analysis) and thence to a polydimethylacrylamide–kieselguhr composite, all ready-made from a commercial source
$N(\alpha)$-protection	Fmoc
Side-chain protection	Lys(Boc), Ser(Bu^t), Thr(Bu^t), Tyr(Bu^t), Gln(Trt), His(Trt), Arg(Mtr)
$N(\alpha)$-deprotection	20% piperidine/DMF
Coupling	FmocXaaOPfp/DMF in most cases, except after an Hmb-residue, when an Fmoc NCA/CH$_2$Cl$_2$ was used to add the next residue

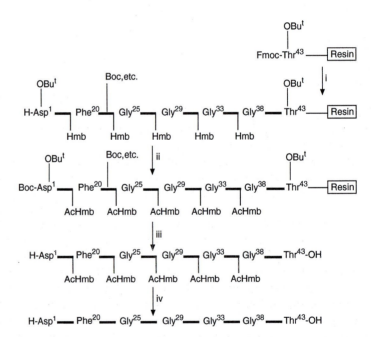

Scheme 9.10.   Stages: i, chain assembly (see Table 9.2); ii, Boc$_2$O, then Ac$_2$O-DIPEA; III, TFA, then purification of the soluble derivatized peptide which is released; iv, NH$_2$NH$_2$/DMF, then TFA.

Conclusion 9.4

So what is the best way of making a peptide? The history of the subject shows that any generalization is likely to be a hostage to fortune, but the answer seems to be at present that solution methods are best for very small peptides , solid phase methods win for intermediate sizes (with some version of the Sheppard approach having the popular vote), and the way for proteins is solution assembly of fragments made by solid phase procedures.

10 GFP, a *tour de force*

The green fluorescent protein (GFP) of the jellyfish *Aequorea victoria* is a single chain protein of 238 residues. The fluorophore is formed by a unique and incompletely understood backbone modification of the region -Ser^{65}-Tyr^{66}-Gly^{67}—which is spontaneous when conditions are right.

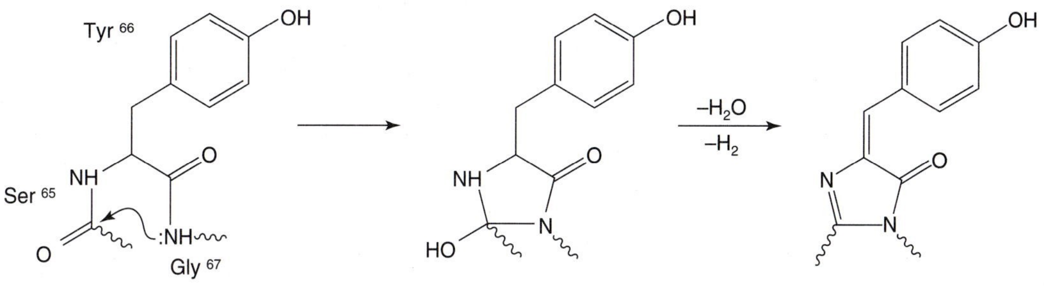

Scheme 10.1   The spontaneous generation of the GFP fluorophore.

The gene for the sequence can be made to express itself in other organisms, and GFP is being used as a visible marker in wide range of molecular biology and biomedical work.

At the time its synthesis was undertaken by Sakakibara and his team, GFP was some hundred residues longer than the longest sequence which had been prepared chemically , and it is still at the time of writing the biggest protein to have been synthesized. That obviously makes it a case of particular interest. By the yardsticks of classical organic chemistry, the yields in the final stages were not good after purification, and characterization was not complete, but those would be petty criticisms of a monumental achievement. It involved over 500 separate chemical reactions starting from protected amino acids. But it is more a monument to technical skill, patience and labour than to novel chemistry—the chemical procedures used were at several points subtle refinements of usual practice, but do not involve new principles. The general strategy adopted had been worked out on much smaller molecules. The overall plan of attack was one based on fragment condensation in solution with almost maximal relatively stable side-chain 'permanent' protection (Table 10.1) in conjunction with Boc α-amino 'temporary' protection. Short protected fragments of about 10 residues each, mostly chosen to have *C*-terminal Gly or Pro, were prepared by stepwise solid phase peptide synthesis (SPPS; Table 10.2); these short fragments were assembled on four fronts to give four major fragments of 50–60 residues which were coupled in pairs to give two fragments of roughly equal size which were then combined and deprotected (Scheme 10.2).

Table 10.1   Side-chain protecting groups for GFP synthesis

Xan

Arg	Tos	See p.44
Asn	Xan	
Asp	cHx	See p.45
Cys	Acm	See p.46
Glu	cHx	
Gln	Xan	
His	Bom	See p.50
Lys	Z[2Cl]	See p.71
Ser	Bzl	See p.48
Thr	Bzl	See p.48
Trp	Hoc, cyclohexyloxycarbonyl	See p.51
Tyr	Pen, pent-3-yl	

There were four main considerations behind the choices listed in the above table. *First*, survival of α-Boc acidolysis conditions; all pass that test except Xan, which is only marginally less acid-labile than Boc. But since side reactions involving the amide side-chain take place mainly when the α-carboxyl is activated for coupling (see Schemes 6.18 and 6.19), it did not matter that Xan groups were lost during the next Boc acidolysis after Asn or Gln incorporation. *Secondly*, survival of the basic/nucleophilic conditions for detaching protected fragments after SPPS assembly (see Table 10.2 below): the usual Z(2Br) group for Tyr failed on that, and so the Pen group was devised—similarly, formyl proved unsuitable for Trp and so Hoc was developed. *Thirdly*, survival of phenacyl reductive cleavage conditions: all are satisfactory on that score. *Fourthly and finally*, removability by HF treatment: all except Acm are cleaved by HF, and in that case a subsequent treatment with Hg(OAc)$_2$ was necessary.

Protected peptides anchored through an acid-stable, base-labile ester can be assembled on this carrier by a Merrifield approach (α-Boc, acid resistant side-chain blockade). Base treatment (*cf.* Fmoc cleavage) gives peptide free acids with all *N*-terminal and side-chain protection intact.

Table 10.2   SPPS of short protected GFP fragment acids

Carrier	Polystyrene extended with a base-labile linker -see margin
$N(\alpha)$-protection	Boc
Side-chain protection	See Table 10.1 above
Attachment of the first residue	(BocXaa)$_2$O/DMAP/CH$_2$Cl$_2$
$N(\alpha)$-deprotection	TFA/ CH$_2$Cl$_2$
Coupling	BocXaaOH/HBTU/HOBt/DIPEA *N*-methylpyrrolidone as solvent
Cleavage from the carrier	20% morpholine/DMF

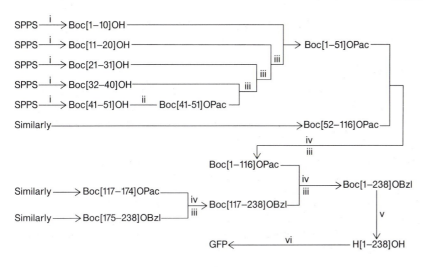

Scheme 10.2 Conditions: i, see Table 10.2; ii, CO_2H protection by reaction with $PhCOCH_2Br$; iii, Boc removal and coupling; iv, CO_2H deprotection with Zn/AcOH; v, total deprotection; vi, spontaneous fluorophore formation.

The short protected peptide acids were all thoroughly purified, and, in cases where the strategy required it, the *C*-terminal carboxy group was converted to a Pac ester by treatment with phenacyl bromide. No His(Bom) residues, which would probably have reacted with phenacyl bromide, were present in these cases. Boc cleavage was with TFA, and WSCI (see p. 32) was used as coupling reagent, with HOOBt as additive instead of the more popular HOBt. As the terminal residue of the carboxy component was Gly or Pro in most couplings, there was no risk of racemisation. Where this was not the case, no difficulty was encountered in practice. The principal problem was solubility, and the the venture turned as much as anything on the fact that mixtures of chloroform and TFE or phenol were good solvents for the protected intermediates and also compatible with the coupling chemistry. For the Pac cleavage stage hexafluoroisopropanol (HFIP) was added to assist dissolution when necessary. About 700 mg of the fully protected complete sequence Boc[1-238]OBzl was obtained. Treatment of 465mg with HF at −5° for an hour in the presence of a similar amount of cysteine and a large excess of cresol as scavengers removed all except the Acm protecting groups and gave an opportunity for purification (gel filtration, and HPLC) and characterization—amino acid analysis and mass spectrometry gave results in accord with theory for H[1-238(Acm$_2$)]OH, and capillary zone electrophoresis was consistent with homogeneity. At this stage there was 33mg in hand. Finally the Acm groups were removed with Hg (OAc)$_2$ in 50% acetic acid, and after gel filtration and treatment under the conditions used for renaturing GFP after denaturation, the synthetic protein solution developed fluorescence characteristics identical to the native protein. Perhaps the most remarkable thing of all, however, is that in reflecting on this and numerous other major syntheses performed by broadly the same approach, Sakakibara said in 1997 'I am certain in general that we are able to synthesize an ordinary 100-

residue protein within several months and a 200-residue protein within a year.'

Sources

It is less than a decade since the structure of GFP was fully worked out and the possibilities for using it as a marker in molecular biology were recognised, but there have already been several books about it and the internet is rich with information.

Phillips, G.N., Jr., 'Structure and dynamics of green fluorescent protein', *Current Opinion in Structural Biology*, 1997, **7**, 821.

Sakakibara, S., *et al.*, 'Chemical synthesis of the precursor molecule of the Aequorea green fluorescent protein, subsequent folding, and development of fluorescence', *Proc. Nat. Acad. Sci. USA*, 1998, **95**, 13549.

Sakakibara, S., 'Chemical synthesis of proteins in solution', *Biopolymers (Peptide Science)*, 1999, **51**, 279.

Appendix A Abbreviations and their use

In 1947, Brand and Edsall set out suggestions for the concise formulation of proteins, using abbreviated designations for the constitutent amino acids. Their system has been developed under the guidance of an International Commission, which made detailed recommendations in 1983.

Each amino acid has a symbol, usually comprising three letters derived from its familiar trivial name, beginning with the initial letter. The common amino acids of proteins also have single-letter symbols, but these were primarily intended for the presentation of sequence data, and are not used in this book, although they are starting to come into use for synthetic work as the scope of peptide synthesis widens to embrace ever more cumbersome structures. Table A.1 lists the standard symbols for the coded amino acids.

For a fuller account see *J. Peptide Science*, 1999, **5**, 465.

Table A.1 Symbols for the common proteinogenic amino acids

Three-letter symbol	Amino acid*	One-letter symbol
Ala	Alanine	A
Arg	Arginine	R
Asn**	Asparagine	N
Asp**	Aspartic acid	D
Cys	Cysteine	C
Gln**	Glutamine	Q
Glu**	Glutamic acid	E
Gly	Glycine	G
His	Histidine	H
Ile	Isoleucine	I
Leu	Leucine	L
Lys	Lysine	K
Met	Methionine	M
Phe	Phenylalanine	F
Pro	Proline	P
Ser	Serine	S
Thr	Threonine	T
Trp	Tryptophan	W
Tyr	Tyrosine	Y
Val	Valine	V

*For structures, see Table 1.1.
**Asx and Glx are used, especially in sequence analysis work, when there is Asp/Asn or Glu/Gln uncertainty.

The essential principle is that the symbols represent whole free amino acids (α-amino nitrogen to the left, α-carboxy to the right, side-chains above or below) when they stand alone, but amino acid residues -NHCHRCO- when substituents or other residues are shown contiguously or attached by bonds; if nothing is indicated at any α- or side-chain position it is understood to be unsubstituted, but H- or -OH may be added for emphasis. Thus lysylglutamic acid (**1**) may be formulated in various ways, including **2–4**, but not **5** or **6**, which represent isomers with different inter-residue amide links.

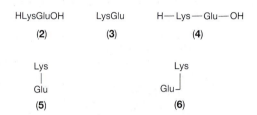

$(CH_2)_4NH_2$

$H_2NCHCONHCHCO_2H$

$(CH_2)_2CO_2H$

(**1**)

HLysGluOH	LysGlu	H—Lys—Glu—OH
(**2**)	(**3**)	(**4**)

Lys	Lys
|	|
Glu	Glu⌐
(**5**)	(**6**)

Derivatives of amino acids and peptides can be represented by combining the amino acid symbols with substituent symbols (Table A.2).

Table A.2   Substituent and derivative symbols*

Symbol	Substituent
Ac	Acetyl
Acm	Acetamidomethyl
Alloc	Allyloxycarbonyl
[Bn]	Benzyl, Bzl
Boc	*t*-Butoxycarbonyl
[tBoc]	*t*-Butoxycarbonyl, Boc
Bom	π-Benzyloxymethyl
Bpoc	2-(4-Biphenylyl)isopropoxycarbonyl
Bu^i	*i*-Butyl
Bu^n	*n*-Butyl
Bu^t	*t*-Butyl
Bz	Benzoyl
[Bzh]	Benzhydryl (= diphenylmethyl, Dpm)
Bzl	Benzyl
[Cbz]	Carbobenzoxy (= benzyloxycarbonyl, Z)
cHx	Cyclohexyl
Dcha	Dicyclohexylammonium salt
Dde	1-(4,4-Dimethyl-2,6-dioxocyclohexylidine)-ethyl
Dnp	2,4-Dinitrophenyl
Dpm	Diphenylmethyl
Dts	Dithiasuccinoyl
Et	Ethyl
Fmoc	9-Fluorenylmethoxycarbonyl
For	Formyl
Hmb	2-Hydroxy-4-methoxybenzyl
Hoc	Cyclohexyloxycarbonyl

Me	Methyl
Mtr	2,3,6-Trimethyl-4-methoxybenzenesulphonyl
Nps	2-Nitrophenylsulphenyl
OBt	Benzotriazol-1-yl ester
ONp	4-Nitrophenyl ester
OPcp	Pentachlorophenyl ester
OPfp	Pentafluorophenyl ester
OSu	Succinimido ester
OTcp	2,4,5-Trichlorophenyl ester
Pac	Phenacyl ($PhCOCH_{2-}$)
Pen	Pent-3-yl
Ph	Phenyl
Pht	Phthaloyl
Scm	Methoxycarbonylsulphenyl
Pmc	2,2,5,7,8-Pentamethylchroman-6-sulphonyl
Pr^i	*i*-Propyl
Tbfmoc	Tetrabenzo-Fmoc
Tfa	Trifluoroacetyl
Tos	4-Toluenesulphonyl
[Tri]	Trityl, Trt
Trt	Trityl (= triphenylmethyl)
Xan	9*H*-Xanthen-9-yl
Z	Benzyloxycarbonyl
Z(2Br)	2-Bromobenzyloxycarbonyl
Z(2Cl)	2-Chlorobenzyloxycarbonyl
Z(OMe)	4-Methoxybenzyloxycarbonyl

*Square brackets indicate alternative abbreviations which are found in the literature.

α-Benzyloxycarbonyl, ε-*t*-butoxycarbonyl-lysine succinimido ester, for example, is **7** or **8**, and α-methyl,γ-benzyl-glutamic acid diester hydrochloride is **9** or **10**; reaction of the two in the presence of base would give the fully protected dipeptide **11** (otherwise **12**). Note that a

ZLys(Boc)OSu

(7)

Z—Lys—OSu
|
Boc

(8)

HCl,HGlu(OBzl)OMe

(9)

HCl,H—Glu—OMe
|
OBzl

(10)

ZLys(Boc)Glu(OBzl)OMe

(11)

Z—Lys—Glu—OMe
|　　|
Boc　OBzl

(12)

substituent symbol, unlike an amino acid residue symbol, may be attached from its right or its left or at any angle from above or below it, without any change in its significance.

It is convenient to define here the conventional abbreviations for common solvents, etc., and also the generalized symbols used in the schemes: see Tables A.3 and A.4 respectively.

Table A.3   Abbreviations for some common reagents and solvents, etc

Abbreviation	Full name
BOP	Benzotriazol-1-yl-oxy-tris-dimethylaminophosphonium PF_6^- salt
DCCI	Dicyclohexylcarbodiimide (also DCC)
DICI	Diisopropylcarbodiimide
DIPEA	Diisopropylethylamine
DMAP	4-Dimethylaminopyridine
DMF	Dimethylformamide
DMS	Dimethylsulphide
DMSO	Dimethylsulphoxide
DPPA	Diphenylphosphoryl azide
EEDQ	2-Ethoxy-1-ethoxycarbonyl-1,2-dihydroquinoline
HATU	O-(7-Azabenzotriazol-1-yl)-N, N, N', N'-tetramethyluronium PF_6^- salt**
HFIP	Hexafluoroisopropanol
HMP	Hexamethyphosphoramide (also HMPA, HMPT)
HOAt	1-Hydroxy-7-azabenzotriazole
HOBt	1-Hydroxybenzotriazole
HOCt	Ethyl 1-hydroxy-1,2,3-triazole-4-carboxylate
HOOBt	3,4-Dihydro-3-hydroxy-4-oxo-1,2,3-benzotriazine (also HODhbt)
NCA	N-Carboxyanhydride
NMM	N-Methylmorpholine
PAM resin	Phenylacetamidomethyl resin
PyBOP	Benzotriazol-1-yl-oxy-tris-pyrrolidinophosphonium PF_6^- salt
TBTU	O-(Benzotriazol-1-yl)-N, N, N', N'-tetramethyluronium BF_4^- salt**
TFA*	Trifluoroacetic acid (CF_3CO_2H)
TFE*	Trifluoroethanol (CF_3CH_2OH)
THF	Tetrahydrofuran
UNCA	Urethane protected NCA
WSCI	Water-soluble carbodiimide 1-ethyl,3-(3′dimethylaminopropyl)—carbodiimide hydrochloride (also WSC, EDC, EDCI)

*There is an exquisite and subtle confusion between some English and German literature over these abbreviations. Throughout this book, in accord with convention, we use TFA for CF_3CO_2H, derived from *TriFluoroAcetic acid* in English, and TFE for CF_3CH_2OH, derived from *TriFluoroEthanol* in English. Derivation from the the German *TriFluorEssigsäure* and *TriFluorÄthanol* would interchange the two abbreviations.
**So-called. HBTU is the PF_6^- salt corresponding to TBTU.

Table A.4   Generalized symbols used in the schemes

Ar	An unspecified aryl substituent
B	An unspecified base
Nu	An unspecified nucleophile
P,P^1.....P^n	Unspecified protecting groups
(PA)	Polyacrylamide-type (Sheppard) resin
(PS)	Polystyrene-type (Merrifield) resin
R,R^1......R^n	Unspecified alkyl groups
X	An unspecified leaving group
Xaa, Xaa^1.....Xaa^n	Unspecified amino acid residues
∿	The rest of the molecule

Appendix B Further reading and general sources

Jones, J. H. (2000). 'The chemical synthesis of peptides: a select bibliography', *J. Peptide Science*, **6**, 201. This bibliography lists and classifies with comment nearly 150 relevant book, key reviews, and symposium proceedings. It was originally compiled for a projected second edition of the author's monograph listed below, which was abandoned because of the appearance of a number of excellent new books covering essentially the same material at a similar advanced level. It is a 'select' bibliography—the field is so large that an exhaustive one would be at risk of self-defeat. But it is also quite wide-ranging and, it is hoped, thorough.

Jones, J. [H.] (1991). *The chemical synthesis of peptides*. Oxford University Press. The first printing of the present Primer was a distillate of this monograph, which may still be found useful for the location of specific references and historical perspective, but it is now somewhat dated.

Barrett, G. C. (ed.) (1985). *Chemistry and biochemistry of the amino acids*. Chapman and Hall, London.

Williams, R. M. (1989). *Synthesis of optically active α-amino acids*. Pergamon Press, Oxford. For an updating review, see Duthaler, R. O. (1994) *Tetrahedron*, **50**, 1539.

Lloyd-Williams, D., Albericio, F., and Giralt, E. (eds.) (1997). *Chemical approaches to the synthesis of peptides and proteins*. CRC Press, Boca Raton, USA.

This elegant more advanced book would be by far the best next step after the present Primer.

Hecht, S. M. (ed.) (1998). *Bioorganic Chemistry. Peptides and Proteins*. OUP.

Chan, W. C. and White, P. D. (eds.) (2000). *Fmoc solid phase peptide synthesis. A practical approach*. OUP.

Andersson, L., Blomberg, L., Flegel, M., Lepsa,, L., Nilsson, B., and Verlander, M. (2000). 'Large-scale synthesis of peptides', *Biopolymers (Peptide Science)*, **55**, 227. This valuable review not only covers its title topic well, but also gives insights into peptide drug development and the peptide drug market generally.

The spirit of what can now be called industrial peptide chemistry is now decidedly upbeat

Loffet, A. (2002). 'Peptides as drugs: is there a market?', *J. Peptide Science*, **8**, 1. As the author makes clear, there most certainly now is a major market for "Nature's pharmaceuticals".

Index

Citations of common protecting groups and reagents are listed here using standard abbreviations (see Appendix A) or formulae.